IC 555 PROJECTS

by

E.A. PARR, B.Sc., C.Eng., M.I.E.E.

BERNARD BABANI (publishing) LTD
The Grampians
Shepherds Bush Road
London W6 7NF
England

PLEASE NOTE

Although every care has been taken with the production of this book to ensure that any projects, designs, modifications and/or programs, etc. contained herein, operate in a correct and safe manner and also that any components specified are normally available in Great Britain, the Publishers do not accept responsibility in any way for the failure, including fault in design, of any project; design, modification or program to work correctly or to cause damage to any other equipment that it may be connected to or used in conjunction with, or in respect of any other damage or injury that may be so caused, nor do the Publishers accept responsibility in any way for the failure to obtain specified components.

Notice is also given that if equipment that is still under warranty is modified in any way or used or connected with home-built equipment then the warranty may be void

This book is dedicated to W. & G.B.

Acknowledgements:
The author would like to thank the Editors of Practical Electronics and Electronics Today International for permission to use material that first appeared under the author's name in those magazines.

The author would also like to thank Mullard Ltd for permission to use data on the 555.

Finally there is my wife who did all the typing against great odds.

E. A. PARR

ISBN 0 85934 047 3
First published – February 1978
Revised and Reprinted – October 1979
Reprinted – September 1980
Revised and Reprinted – November 1981
Reprinted – April 1985
Reprinted – December 1986
Reprinted – July 1988
Reprinted – June 1989
Reprinted – April 1990
Reprinted – December 1991
Reprinted – January 1993
Reprinted – October 1994
Reprinted – October 1996
Reprinted – January 1998
Reprinted – January 2000
Reprinted – June 2001
Reprinted – June 2002
Reprinted – June 2003
Reprinted – May 2004
Reprinted – May 2005
Reprinted – May 2006
Reprinted – Feb 2008
Reprinted – Jan 2010

Cover Design by Gregor Arthur
Printed and Bound in Great Britain by Cox & Wyman Ltd, Reading

CONTENTS

1. The 555 Timer

Every so often a device appears that is so useful that one wonders how life went on before without it. The 555 timer is such a device.

It was first introduced by Signetics, but is now manufactured by almost every semiconductor manufacturer. It is cheap and probably the most versatile of the three devices we shall describe.

The device can be made to operate as a monstable or as an oscillator (multivibrator) with times from micro seconds to several hours.

It can operate on power supplies from 5v to 18v and can thus be used with TTL, model railways or motor car circuits with ease.

Finally, and not least, the device can source or sink up to 200mA (0.2A) allowing it to drive relays, lamps and other large loads directly.

The 555 itself takes about 10mA from the supply when the output is high (timed period) and ImA in the reset state (output low). To this must be added the load current.

2. Basic Circuits

This section describes the basic circuits that can be built around the 555 timer. These will be used in specific applications in later sections.

2:1 Monostable

The basic 555 is usually obtained in an 8 lead D.I.L. pack with connections and internal circuit as shown in Fig. 2:1.

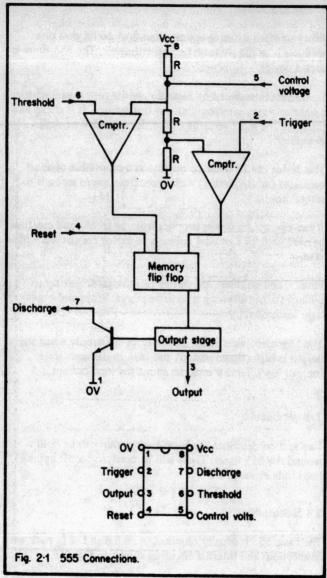

Fig. 2·1 555 Connections.

Vcc (the supply) and ground (Ov) are obvious. The output is true (high going) and the trigger input is low going. The device is D.C. coupled, hence the output stays up if the trigger input is longer than the timed period. The device is not re-triggerable in its basic form.

To construct a monostable we add a resistor and capacitor Rt and Ct as shown in Fig. 2:2.

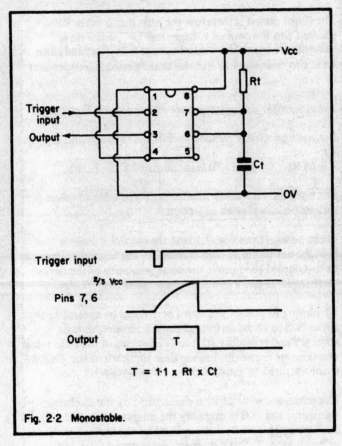

$$T = 1.1 \times Rt \times Ct$$

Fig. 2·2 Monostable.

9

Referring to Fig. 2:1 and 2:2 the operation is as follows. The trigger input sets the flip flop and the output goes high. The discharge transistor turns off, and Ct starts to charge via Rt. When the voltage on Ct reaches the control voltage defined by the three resistor chain, comparator 1 resets the flip flop, the output goes low and the discharge transistor turns on again to discharge Ct. The circuit can now be triggered again by another input.

The timed period is therefore the time that it takes Rt to charge Ct to the control voltage from 0v. As the three resistors are equal, the control voltage is $\frac{2}{3}$ Vcc, and since Rt is also connected to Vcc the timed period is independent of Vcc.

Waveforms for the operation are shown also on Fig. 2:2

For mathematically minded people, the timed period is given by:—
$T = I \cdot I \, Rt \times Ct$ where Rt is in ohms and Ct in Farads.

For non mathematically minded people, a table of values for various periods is given in section 5.

It can be seen from Fig. 2:1 that the control voltage is brought out to pin 5. With this facility the control voltage can be de-coupled to improve the noise immunity of the device, or changed to give a control voltage other than $\frac{2}{3}$ Vcc.

By varying Rt and Ct the timed period can be controlled from about 5uS to about an hour. Above 5 minutes, though, accuracy and reliability start to fall because of the large value components necessary. In long time applications the ZN1034 (manufactured by Ferranti:) is a better device.

The minimum value of Rt is determined by the discharge transistor, and IKO is normally the minimum allowed. The maximum value of Rt is determined by the leakage current of comparator 1. The data sheets recommend 20M, but accuracy suffers above IMO. For most purposes Rt should be kept between IKO and IMO.

There are no limits of Ct apart from cost. Note that large value electrolytic capacitors have high leakage currents which can cause large variations away from the calculated time periods. If very large values are used, the discharge transistor may take an appreciable time to discharge Ct.

If electrolytic capacitors are used, the voltage rating should be about the same as the supply Vcc. An electrolytic capacitor does not become a capacitor until about O.I of its voltage rating. If, say, 100v rated capacitor was used on a 15 volt supply with a 555 we would see wave forms similar to Fig. 2:3

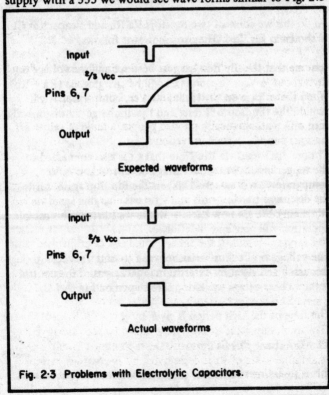

Fig. 2·3 Problems with Electrolytic Capacitors.

which are much shorter than calculations would lead us to expect.

Obviously the voltage rating of the capacitor should not be below Vcc or the rapid demise of the capacitor will result. The mess left behind is not pleasant to clean off the board.

2:2. Multivibrator (Astable Operation)

If we look at Fig. 2:1 we see that the trigger input is connected to the flip flop via comparator 2. To set the flip flop the input has to fall below $\frac{1}{3}$ Vcc. Apart from providing noise immunity, this allows us to make the 555 into an oscillator.

To do this we connect two resistors Ra,Rb and a capacitor Ct as shown in Fig. 2:4. The operation is as follows.

Assume that the flip flop has just been set and the voltage on Ct is about $\frac{1}{2}$ Vcc. Capacitor Ct will be charging via (Ra + Rb). When the voltage on Ct reaches $\frac{2}{3}$ Vcc, comparator 1 will switch, the flip flop will reset and the discharge transistor will turn on.

Ct now discharges via Rb. Note that Ct is also connected to the trigger input. When the voltage on Ct reaches $\frac{1}{3}$ Vcc, comparator 2 will switch. This sets the flip flop again, turns the discharge transistor off and Ct starts charging again via (Ra + Rb). We are now back to where we started. This astable operation will continue indefinitely.

The voltage on Ct thus varies between $\frac{1}{3}$ and $\frac{2}{3}$ Vcc, although this can be changed by external manipulations of the control voltage. The voltage waveforms are shown on Fig. 2:4.

The time of the high period is given by

$$T1 = 0\cdot7 \ (Ra + Rb) \ Ct$$

this is sometimes called the charge period for obvious reasons.

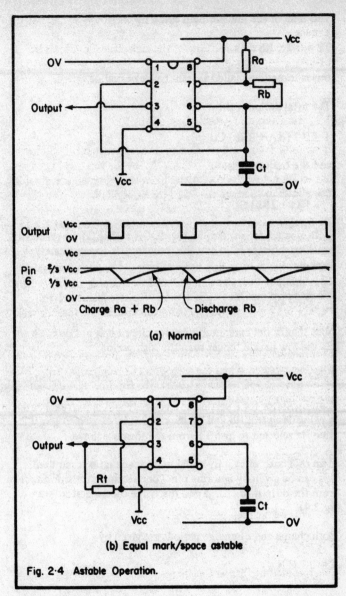

(a) Normal

(b) Equal mark/space astable

Fig. 2·4 Astable Operation.

13

The time of the low period is given by

$$T2 = 0.7 \times Rb \times Ct$$

this is sometimes called the discharge period.

The total period is thus:—

$$T = 0.7 (RA + 2Rb) Ct$$

and the frequency is:—

$$f = \frac{1.45}{(Ra + 2Rb) Ct}$$

If Rb is much larger than Ra (eg. IOOK and IKO) we get nearly equal mark/space ratio and the frequency is approximately given by:—

$$f = \frac{0.72}{Rb \times Ct}$$

Note that in the basic multivibrator circuit it is not possible to get the low period longer than the high period.

Tables of values for various frequencies are given in section 8.

If an oscillator for a particular frequency and mark/space ratio is being designed, Rb and Ct should first be chosen to give T2, then Ra selected to give T1 from the formula above.

If an oscillator of exactly equal mark space ratio is required, this can be given by ignoring pin 7 connecting a resistor direct from the output to charge and discharge the capacitor (see fig 2.4).

Both charge and discharge periods are given by

$$T = 0.7 C R$$

hence the total period is

$$T = 1 \cdot 4 \, C \, R$$

2:3. Edge Triggering and Monostable Chains

In section 2:1 we saw that the 555 is D.C. coupled. If the input is held low beyond the timed period the output stays high until the input goes back high.

To make the 555 into an edge triggered device we need an additional capacitor, resistor and diode (R1, C1, D1) as shown on Fig. 2:5.

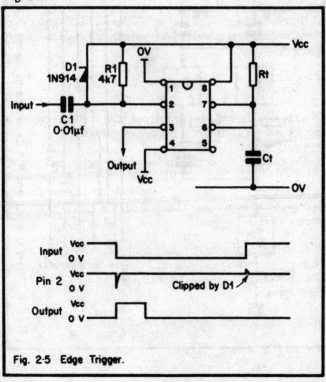

Fig. 2·5 Edge Trigger.

This small network differentiates the input signal to give a small pulse to trigger the 555. The values shown will be adequate for all timed periods.

With an RCD network on several 555's we can make a chain of timers each triggering the next as shown in Fig 2:6.

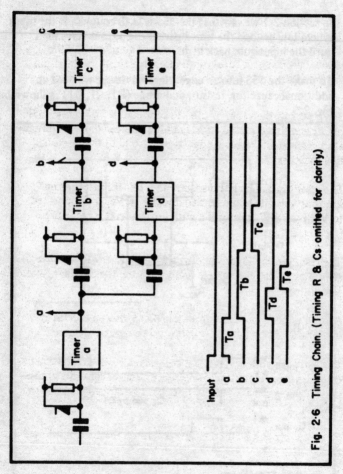

Fig. 2·6 Timing Chain. (Timing R & Cs·omitted for clarity)

Similar monostable chains can be used for sequence controls or test procedures. As can be seen in the example on Fig. 2:6 the chain can branch and one 555 can trigger several others.

2:4. Reset Facility

Pin 4 on the 555 is called the reset. In all the circuits up to now this pin has been tied to Vcc. If the reset pin is taken to 0v, the internal flip flop is reset, the output goes low and the discharge transistor turns on to discharge Ct. If the reset pin is taken back to Vcc the 555 will remain in its reset state.

The reset can therefore be used to terminate a 555 monostable before the timed period ends, inhibit a 555 oscillator or terminate a monostable chain similar to Fig. 2:6 if all the resets are tied together.

The reset pin can be left disconnected, but for safety from spurious noise it should be tied directly to Vcc, or connected to Vcc by a 4K7 resistor if a push button reset is to be used. (see Fig. 2:7.)

The reset pin can be driven directly off logic or other monostables providing the voltage levels from the logic are similar to Vcc and 0v.

As well as being used as a reset, pin 4 can also be used as an inhibit. If pin 4 is kept at 0v, the 555 cannot be triggered.

In a start up sequence, say, the continuance of the sequence could be made conditional on some external event by connecting a suitable signal onto the reset pin of a strategic 555 in the chain.

The reset pin can be used to give a turn on reset as also shown on figure 2:7.

17

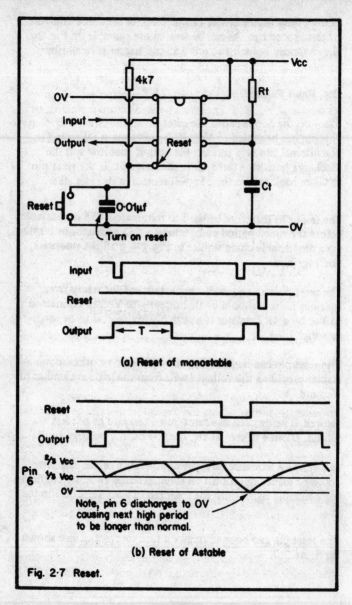

(a) Reset of monostable

(b) Reset of Astable

Note, pin 6 discharges to 0V causing next high period to be longer than normal.

Fig. 2·7 Reset.

2:5. Driving High Current Loads

The 555 can source or sink 0·2A into loads. This can be
extended to 5A by using additional power transistors.

The simplest way is to add emitter follower power transistors
as shown on Fig. 2:8. These can be made current sinking, or
sourcing, or both as required.

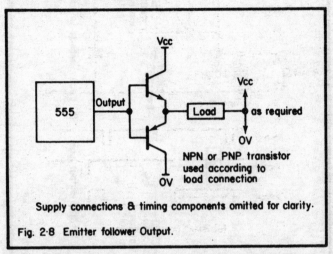

Supply connections & timing components omitted for clarity.

Fig. 2·8 Emitter follower Output.

The one shortcoming of this method is that under large
currents the voltage across the transistors can rise to about 3v
giving a large amount of heat to be dissipated in suitable
heatsinks.

A slightly more efficient way is to use grounded emitter
transistors as shown in Fig. 2:9. The saturation voltage Vcc will
be around 1v giving a correspondingly smaller dissipation. With
the values shown, a 2N3055 will switch around 3 Amps.

If large loads are being switched it is good practice to use
separate logic and load supplies, and to separate 0v connections,
connecting them only at a central earth point. (see Fig. 2:10)

19

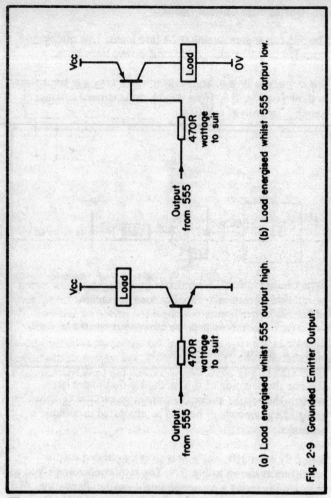

(a) Load energised whilst 555 output high. (b) Load energised whilst 555 output low.

Fig. 2·9 Grounded Emitter Output.

This must be done because wires have a significant inductance and large currents being turned on and off will generate spikes which could cause false triggering of the 555 or even damage in extreme cases.

If the load is inductive, special precautions must be taken as described in section 2:1.

20

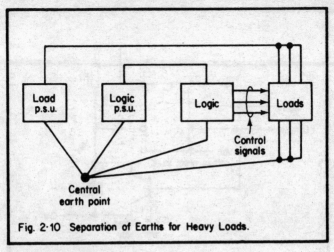

Fig. 2·10 Separation of Earths for Heavy Loads.

2:6. Re-triggerable Circuit

Once triggered, the 555 continues on its way until the timed
period runs out. In many applications it would be useful for
additional trigger inputs to extend the period as required.

Referring back to Figs. 2:1 and 2:2 we see that the timed
period ceases when Ct charges to $\frac{2}{3}$ Vcc. If we can discharge
Ct on each trigger pulse we will extend the timed period.

The simplest way to do this is with an additional transistor TR1,
connected as an emitter follower as shown on Fig. 2:11. Each
input pulse will discharge Ct to about 1 volt. In normal
operation Ct is discharged to about 0·2 volts, so the retriggered
period will be slightly shorter than the normal timed period.
In addition Vbeo of TR1 should be greater than Vcc. A closer
match is obtained by using two transistors as shown in Fig.2:12.
TR2 now discharges Ct to 0·2 volts for each input pulse. The
re-triggered period is the same as the normal timed period. TR1
and TR2 can be any G.P. silicon transistors.

Waveforms for the two re-triggerable variations are shown with
the circuit diagrams.

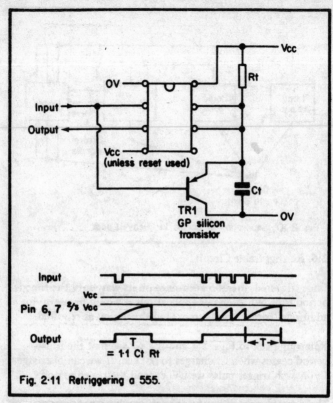

Fig. 2·11 Retriggering a 555.

A re-triggerable monostable can be used to detect the end of a chain of pulses or the failure of an oscillator as shown in Fig. 2:13. Circuits similar to this called Watch Dog Timers are used in computers to detect failures of the computer central processor.

2:7. High Stability Multivibrator

The stability of the period of the 555 depends on the stability of the external R and C and on the stability of the device itself.

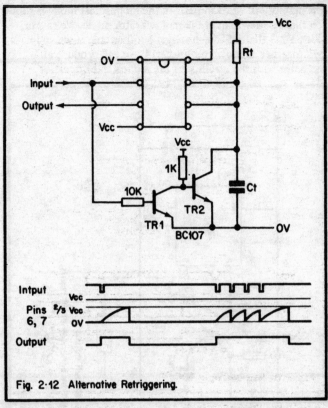

Fig. 2·12   Alternative Retriggering.

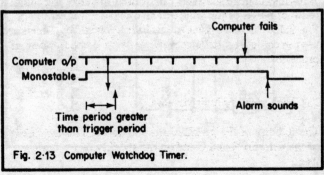

Fig. 2·13   Computer Watchdog Timer.

23

The stability of the 555 itself in the monostable mode is about ten times better than in the multivibrator mode. We saw in section 2:3 that a 555 can trigger another one. If we couple two together as shown on Fig. 2:14, we will produce an oscillator with the stability of the monostable circuit.

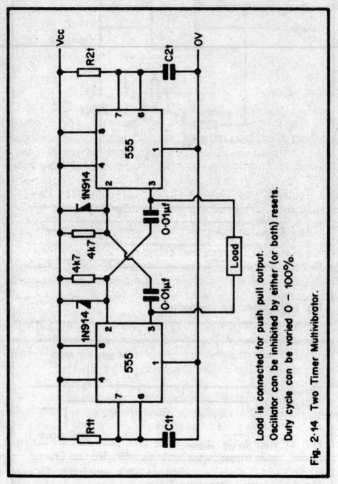

Load is connected for push pull output.
Oscillator can be inhibited by either (or both) resets.
Duty cycle can be varied 0 – 100%.

Fig. 2·14 Two Timer Multivibrator.

24

An additional advantage of this arrangement is that a load connected between the two outputs will be driven in push pull, with current passing through the load in both directions.

If R1 equals R2 (R) and C1 equals C2 (C) the period is given by:—

$$T = 2 \cdot 2R \; C \qquad (1 \cdot 1 \; (R_1 C_1 + R_2 C_2))$$

and the frequency by:—

$$f = \frac{0 \cdot 45}{RC} \qquad \frac{0 \cdot 9}{(R_1 C_1 + R_2 C_2)}$$

2:8. Very Long Time Delays

As mentioned previously, the maximum period obtainable with the 555 is of the order of a few minutes. To obtain longer delays it is necessary to use high value low leakage capacitors. These are expensive, and the accuracy of the delay is poor as the circuit becomes somewhat prone to noise on pin 5.

With additional I.C. s long time delays can be produced at lower cost and improved accuracy. The additional I.C.s are a 7400 quad two input NAND and as many 7490 decade counters as required.

Basically the circuit uses the 555 as an oscillator running at a convenient frequency. The 7490 is a decade counter which devides its input frequency by ten. In the circuit described, two 7490's are used, but the counter chain can be extended indefinitely to give delays of hours, days or even years. The 7400 is used as control logic to start the counter and stop it again when it reaches a predetermined count.

The circuit is shown on Fig. 2:15. IC1 is a 555 connected as an oscillator, with frequency determined by Ra,Rb and Ct as usual. 1C3 and 1C4 are two 7490 counters, with inputs on pin 14 and outputs on pin 11. The link pin 1 is a necessary

25

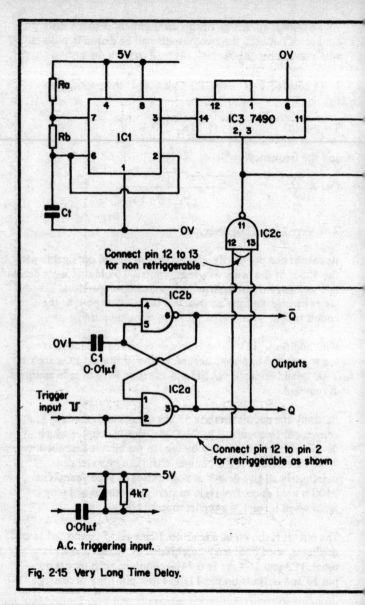

Connect pin 12 to 13 for non retriggerable

Outputs

Trigger input 𝚄

Connect pin 12 to pin 2 for retriggerable as shown

A.C. triggering input.

Fig. 2·15 Very Long Time Delay.

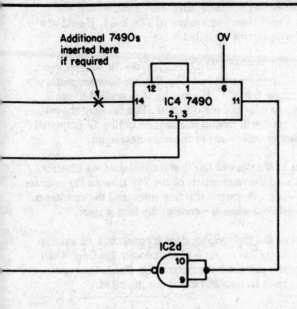

Additional 7490s
inserted here
if required

OV

12 1 6

14 IC4 7490 11

2, 3

IC2d

10

8

9

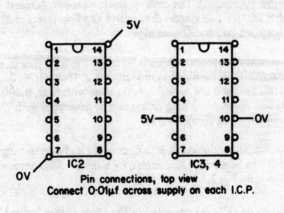

5V

1 14
2 13
3 12
4 11
5 10
6 9
7 8

IC2

OV

1 14
2 13
3 12
4 11
5V 5 10 OV
6 9
7 8

IC3, 4

Pin connections, top view
Connect 0·01µf across supply on each I.C.P.

internal connection in the counter. Pins 2 and 3 reset the counter to a zero state when taken to a logical 1. If held at a logical 1 the counter is inhibited.

The 7400 quad 2 input NAND (1C2) forms the control logic. 1C2a and 1C2b form a set reset flip flop. In the reset state, pin 3 is a '0' and pin 6 is a '1'. If a logical '0' is applied to pin 2, pin 3 will go to a '1' and pin 6 to a '0'. This is called the set state. The circuit will remain in this state until a '0' is applied to pin 4 when it will return to the reset state again.

The output of the control flip flop is connected via a buffer NAND 1C2c to the reset inputs of the 7490, hence the counter is allowed when the control flip flop is set, and the counter is reset and inhibited when the control flip flop is reset.

The output of the top counter stage is connected via another buffer NAND (1C2d) to pin 4 of the control flip flop. When the top counter reaches 8 the control flip flop is thus reset, the counter reset to zero and the count inhibited.

The operation is thus as follows. A low going start pulse is applied to the circuit. 1C2a goes high (Qoutput) and 1C2b goes low ($\bar{Q}$ output). The counter starts to count. When it reaches 80 (for two stages) the control flip flop resets and Q goes low again and $\bar{Q}$ goes high.

The period of the monostable is 80 times the total period of the oscillator, determined as in section 2:2. If three 7490 counters are used the period is 800 times, with four, 8,000 times and so on. Very long delays can be generated with convenient value components.

As shown, the monostable is D.C. coupled and non re-triggerable. It can be made A.C. coupled by a simple RCD network as shown on Fig 2:15.

If pin 12 of 1C2c is connected to the trigger input instead of pin 13 the monostable now becomes re-triggerable. Every time an input pulse occurs the counter resets to zero and starts again.

Capacitor C1 ensures that on first switch on of power, the circuit goes into a reset state. Additional decoupling capacitors should be placed across the supply pins of 1C2 and all the 7490 counters to remove power supply noise.

2:9. Control Voltage Pin 5

In section 2:1 we saw that the control voltage (which determines monostable period and astable frequency) is brought out to pin 5. This was primarily done to allow a capacitor to be added for decoupling in noisy environments.

If, however, the control voltage is varied by external means the monostable period and astable frequency can be varied. The higher the control voltage the longer the monostable period and the lower the astable frequency; the lower the control voltage the shorter the monostable period and the higher the astable frequency.

The three resistors R in Fig. 2:1 are each 5K, hence the control voltage can be varied by an external 5K potentiometer giving an advantage that is not immediately obvious.

If it is desired to alter the period of a monostable (or the frequency of the astable) remotely, the obvious way is to vary the R in the RC timing network with a potentiometer. This means, however, that the charging waveform comes to the potentiometer, and if this is further away from the 555 than a few centimetres the circuit becomes noise prone.

With the connection in Fig. 2:16, the control voltage is varied, and because this is a reference voltage it can be heavily decoupled to remove noise. Control of the 555 with a potentiometer several metres away is quite feasible. This arrangement is advised for adjusting the period (or frequency) with an external potentiometer.

29

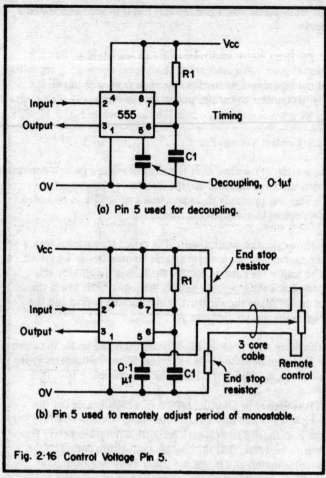

(a) Pin 5 used for decoupling.

(b) Pin 5 used to remotely adjust period of monostable.

Fig. 2·16 Control Voltage Pin 5.

The mathematics relating to the period (and frequency) to control voltage are complex because the charging of the RC network is exponential. If constant current charging (see section 2:11) is used the relationship between control voltage and period becomes relatively linear.

The control voltage can be varied from 0·45 to 0·9 of Vcc in the monostable mode and from 1·7 to 0·9 Vcc in the astable mode.

2:10. Pulse Width and Pulse Position Modulation

If the input to a monostable is driven from a constant frequency clock, and a modulating waveform is applied to pin 5, we get pulse width modulations as shown in Fig. 2:17. The

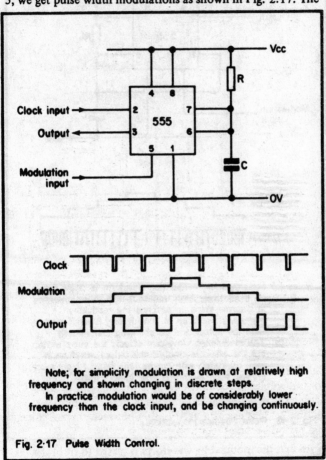

Note; for simplicity modulation is drawn at relatively high frequency and shown changing in discrete steps.
In practice modulation would be of considerably lower frequency than the clock input, and be changing continuously.

Fig. 2·17 Pulse Width Control.

31

modulation is non linear, but it can be made linear by using the constant current source described in section 2:11.

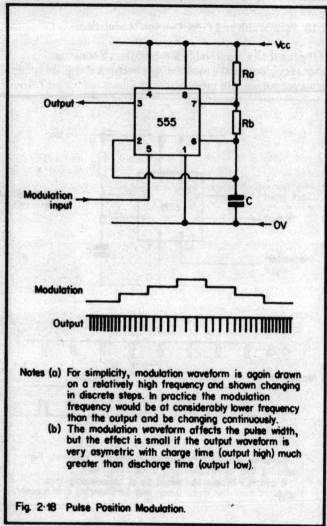

Notes (a) For simplicity, modulation waveform is again drawn on a relatively high frequency and shown changing in discrete steps. In practice the modulation frequency would be at considerably lower frequency than the output and be changing continuously.
   (b) The modulation waveform affects the pulse width, but the effect is small if the output waveform is very asymetric with charge time (output high) much greater than discharge time (output low).

Fig. 2·18 Pulse Position Modulation.

Note that the modulating waveform is biased to fit in the voltage range above.

If a 555 is connected as a highly assymetric astable, and a modulation waveform is applied to pin 5 the modulating wave form varies the high period. This gives the effect of pulse positions modulations. (see Fig. 2:18)

The modulating waveform does, of course, vary the pulse width, but the effect is negligible if Ra is made considerably greater than Rb.

2:11. Constant Current Charging

As described above, the effect of varying the control voltage is non linear. It can be made linear by replacing the timing resistor with a constant current source, as shown in Fig. 2:19.

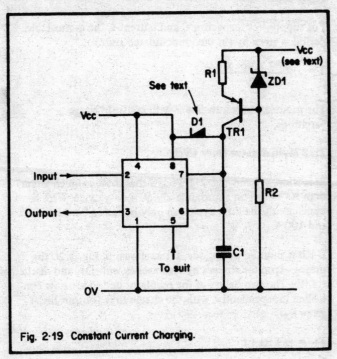

Fig. 2·19 Constant Current Charging.

Transistor TR1 acts as an emitter follower and with zener ZD1 fixes the voltage across R1 and hence the current through R1 and TR1.

The current is given by:—

$$I = \frac{Vz - 0.8}{R1}$$

Remember in the design that the voltage on pin 6 and 7 will rise to $\frac{2}{3}$ of Vcc and this will place a restriction on the zener voltage. The circuit is not practical with a 5v rail, and in this case R1 and ZD1 should be returned to a higher voltage. Diode D1 should then be added to protect the 555 should other components fail and cause pins 7 and 6 to be pulled above the 5v rail.

For supply Vcc, capacitor C and current I, the monostable period is given by (in the unmodulated state)

$$T = \frac{0.67 \times Vcc \times C}{I}$$

The modulation is now linear with control voltage variations.

2:12 Multivibrator Duty Cycle

In section 2:2 and Fig. 2:4 we saw that if Rb is much larger than Ra we get an approximately 50% duty cycle. With the basic circuit the duty cycle can only be varied between 50% and 100%.

If Rb is shunted by a diode, D1 as shown in Fig. 2:20 the timing capacitor charges by Ra in series with D1, and discharges via Rb. The two halves of the oscillator period are now controlled independently, with the charge time (output high) given by:—

$$T1 = 0.7 \, Ra \, Ct$$

34

and the discharge time (output low) given by :

$$T2 = 0.7 \, Rb \, Ct$$

The formula for T1 is only approximate as it ignores the series resistance and voltage drop of D1. For most practical purposes it is accurate enough.

The formula for T2 is precise and is not affected by D1.

With this circuit, duty cycles of less than 50% can be obtained, and the duration of each part of the waveform can be independently controlled.

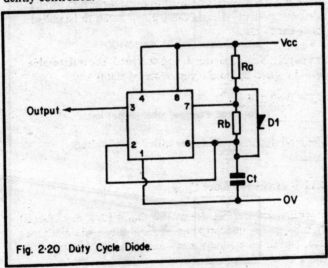

Fig. 2·20 Duty Cycle Diode.

2:13. Inductive Loads

The voltage across an inductor is proportional to the rate of change of current. The 555 timer switches state in about 100 nanoseconds. Very high voltages can thus be generated across inductive loads such as relays, speaker coils or solenoids. These high voltages can cause mistriggering or even damage.

In general, protection is only required for the turn off of the current, and fortunately protection can simply be provided by spike suppression diodes. (also called free wheeling diodes).

For a load connected from Vcc to the 555 the diode is simply connected across the load, cathode to Vcc. When the current is

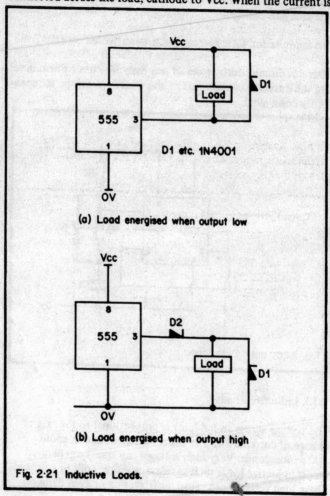

(a) Load energised when output low

(b) Load energised when output high

Fig. 2·21 Inductive Loads.

removed from the load, the inductance causes the 555 end of the load to swing positive, but the diode limits the voltage to 0·8v above Vcc.

For a load connected from the 555 to Ov, two diodes are required. One is a spike suppression diode across the load as before. This limits the voltage to 0·8v below 0v. Unfortunately even this can cause mistriggering, and another diode must be added in series with the load to prevent the 555 output from going below 0v. For some reason this fact is often omitted from 555 data sheets from certain manufacturers, but it is very, very necessary.

These spike suppression diodes are shown on Fig 2:21.

If high current transistors are used as in section 2:5 spike suppression diodes across the load (in the appropriate direction) are all that is necessary.

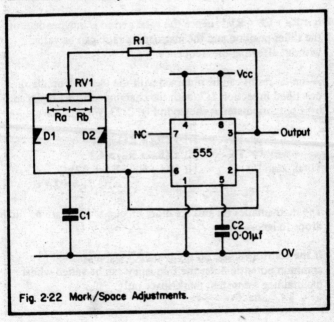

Fig. 2·22 Mark/Space Adjustments.

37

2:14 Mark/Space Adjustment

The circuit allows adjustment of the mark/space ratio of a 555 astable without affecting the frequency. A typical application is a transmitter for a positive control system using pulse width modulation, or as a motor speed control.

The circuit is a variation on the equal mark/space astable in section 2:2. (Fig 2:4b) The circuit with mark/space adjustment is shown on Fig. 2:22. As before the charging and discharging of the timing capacitor is done from the output pin 3, but the output is connected via R1 to the slider of RV1. RV1 is thus split into two halves Ra and Rb. Ra charges C via D1 and Rb discharges C via D2.

The charge period, Ta,	=	$0.7 (Ra + R1) \times C$
The discharge period, Tb,	=	$0.7 (Rb + R1) \times C$
Total period	=	$0.7 (Ra + Rb + 2R1) \times C$

but Ra + Rb = RV1, hence the total period is independent of the slider position and the mark/space ratio can be varied without affecting the frequency.

A similar result can be obtained with the two 555 oscillator described in section 2:7. with the charging resistors replaced by a potentiometer as shown in Fig. 2:23. We now have:—

$$Ta = 1.1 \times (Ra + R1) \times C1$$
$$Tb = 1.1 \times (Rb + R2) \times C2$$
$$\text{Total period} = 1.1 (Ra + Rb + R1 + R2) C$$
$$\text{if } C1 = C2 = C$$

The mathematics for this are more precise as there are no diode drops to ignore.

If the control voltage on the two 555s are connected to a common potentiometer the frequency can be varied whilst maintaining a constant mark/space ratio.

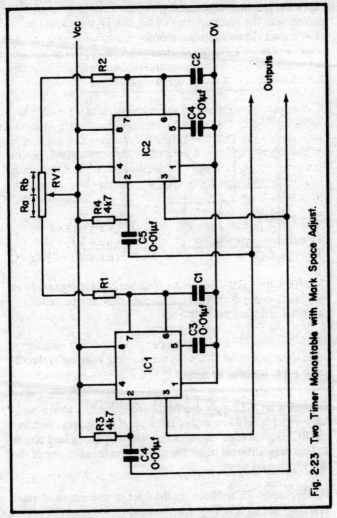

Fig. 2·23 Two Timer Monostable with Mark Space Adjust.

2:15. Initiation by Supply. (Intentional and Unintentional)

The 555 can be triggered on the application of the supply by connecting a capacitor from pin 2 to ground as shown in

Fig. 2:24. The time constant Ra, Ca should be long by comparison with the ramp up time of the supply, but should be shorter than the monostable period.

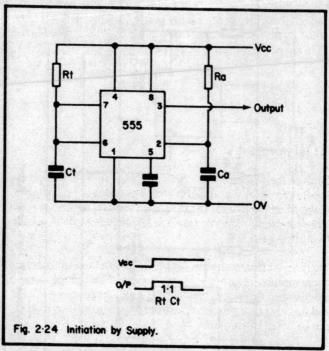

Fig. 2·24 Initiation by Supply.

Triggering on the supply can be caused in circuits where the input pin 2 is held low during ramp up of the supply, such as the RC coupled edge trigger or circuits with a long lead to pin 2. This unintentional triggering can be prevented in any of the three following ways.

Firstly, use an RC network on the reset to give a turn on reset as shown in Fig. 2.7. The time constant of the reset RC network should be long by comparison with the ramp up.

Secondly use a large value capacitor on pin 5 to hold the control voltage down. This keeps the timer reset, as the input is always higher than the voltage on the $\frac{1}{3}$ Vcc comparator (see Fig. 2:1 for internal details of 555).

Finally, place a capacitor across the input pull up resistor to pull pin 2 positive as the supply ramps up. This method is best suited to circuits with long trigger leads and is not suitable for RC coupled edge triggered circuits.

3. Motor Car Circuits

3:1 Introduction

Because the 555 uses any supply rail between 5 and 18v, it is ideally suited for use in motor cars. The following circuits are designed and drawn for negative earth vehicles which are now almost universal. In most cases the circuits are easily adaptible to positive earth.

There are as many schemes for accessories such as indicators, wiper motors etc. as there are motor cars. The circuits described in this section should therefore be taken as guides and not necessarily exact descriptions as to how the circuit can be fitted to your car.

Motor car supplies are notoriously noisy (mainly caused by dynamo or alternator brush noise). If this presents any problems, the circuit should be de-coupled with a large value electrolytic in parallel with a $2u$F polyester capacitor between the supply and Ov at the board. Alternatively a small 9v regulator can be made to give a stabilised 9v rail for the 555.

Heat can present a problem under a motor car bonnet, and the circuits should be kept clear of the engine, radiator and exhaust pipes for obvious reasons.

Needless to say, all circuits should be fused, a battery can deliver a massive fault current.

41

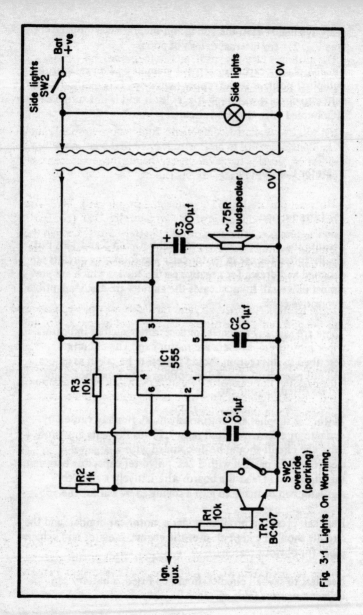

Fig. 3-1 Lights On Warning.

42

3:2. Lights on Warning

The author is very forgetful, and in foggy weather often forgets to turn off the car lights after parking. This circuit gives an audible alarm if the lights are on with the ignition off. The circuit is shown in Fig. 3:1.

IC1 and its associated components form a conventional alarm circuit as described in section 6:2. Note that Vcc comes direct off the side lights on the car, hence the circuit can only oscillate when the lights are turned on.

The reset pin on the 555 is connected to transistor TR1. The base of TR1 is connected via R1 to the 1GN AUX terminal on the fuse box. This terminal is at battery positive when the ignition is turned on, hence TR1 will be turned on when the ignition is on, holding the reset of the oscillator at Ov. This inhibits the alarm when the ignition is on.

If the ignition is turned off, and the lights are on, Vcc is applied to the 555 and TR1 is turned off. The reset is now at Vcc and the alarm starts.

Switch SW1 is optional if it is required to override the circuit for parking. For cars with a "parking" switch which disconnects nearside lights this override can be omitted if Vcc is obtained from the nearside lights.

The simple oscillator used above can, of course, be replaced by any of the more esoteric oscillators available described in section 6.

3:3. Indicator Warning

Not all of us drive Rolls Royces, and in most cars the click of the indicators is not particularly audible. This circuit gives a bleep for each flash of the indicators, and is shown in Fig. 3:2.

43

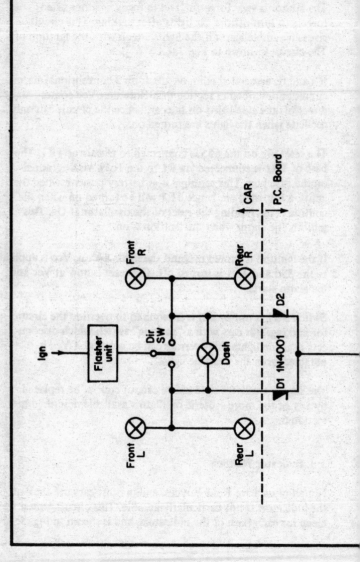

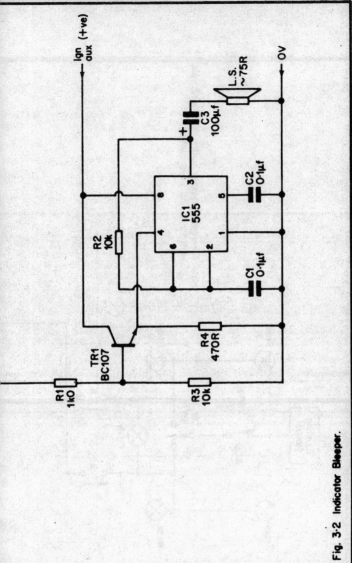

Fig. 3-2 Indicator Bleeper.

45

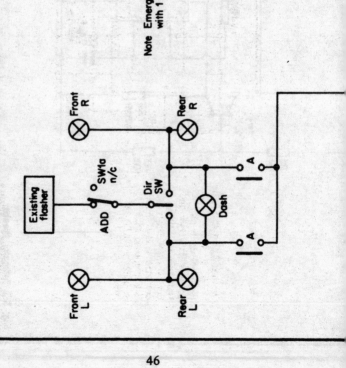

Note Emergency flasher switch SW1 is 2 pole
with 1 n/c and 1 n/o contact.

46

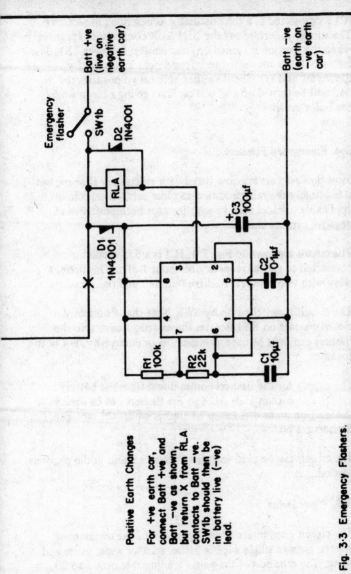

Positive Earth Changes

For +ve earth car, connect Batt +ve and Batt −ve as shown, but return X from RLA contacts to Batt −ve. SW1b should then be in battery live (−ve) lead.

Fig. 3·3 Emergency Flashers.

47

IC1 is connected as a tone oscillator as described in section 6. The supply is derived off the IGN AUX connections (positive when the ignition is turned on) and emitter follower TR1. The base of TR1 is connected to left and right indicators by diodes D1 and D2. Should either bulb set go positive, the 555 will be turned on and will oscillate giving a bleep whilst the bulbs are lit.

3:4. Emergency Flashers

Most modern cars are now fitted with emergency flashers, but it is a highly desirable addition to older cars. As this circuit is applicable to older cars, connections for both positive and negative cars are shown.

The circuit is shown in Fig. 3:3; 1C1 is a 555 oscillator connected to give a frequency of about 1·5Hz. This drives a relay with two contacts to drive both left and right flashers.

The circuit is switched on by SW1. Note that a normally closed contact on SW1 isolates the existing flashers, so the flashers can only be used as indicators or emergency, not both together.

The supply for the flashers comes direct from the battery supply rather than 1GN AUX so the flashers can be used with the ignition on or off. Lamp L1 flashes whilst the emergency flashers are on.

The circuit can be used with the indicator alarm in the previous section.

3:5. Wiper Delay

The circuit gives intermittent operation of the windscreen wipers, giving a single wipe, a pause, another wipe, pause and so on. The duration of the pause is adjustable over 2 to 20 secs.

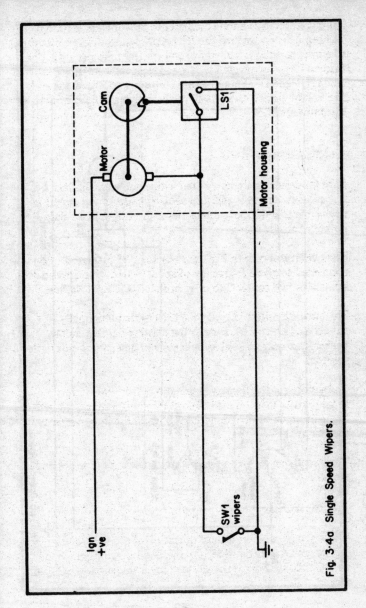

Fig. 3·4a Single Speed Wipers.

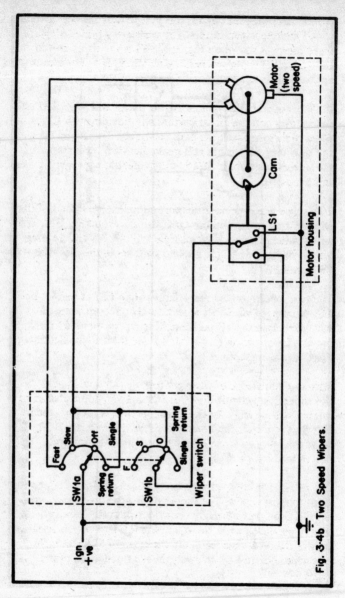

Fig. 3-4b Two Speed Wipers.

50

Before describing the circuit, it is best to describe the operation of self parking wiper motors. The wiring of a typical single speed motor is shown on Fig. 3:4. SW1 is the wiper switch, and LS1 is the wiper self parking switch. This can be a cam operated limit to a brush operated copper plate.

If SW1 is operated momentarily, the motor will start, LS1 will change over and the motor current will flow through LS1. When the wipers reach the park position, LS1 changes back and providing SW1 is not still made the wipers will stop in the correct position. If SW1 is still made, the wipers will make another sweep.

Two speed wipers are a little more complex, the circuit of the wipers fitted to the authors car is also shown on Fig. 3:4. The wiper switch has four positions; Fast, Slow, Off and a spring loaded single wipe. There are two switches in the mechanism SW1a and SW1b.

The two speed motor drives a limit switch LS1 as before, but LS1 is arranged via SW1b so it can only select slow speed when SW1 is in the off position. This arrangement is to prevent LS1 from Selecting "slow" and the switch selecting "fast" at the same time.

With our knowledge of wiper motors we can now look at the wiper delay circuit Fig. 3:5. 1C1 is a 555 connected as an astable, driving relay A. The discharge time is set for 1 sec by R2, C1 and the charge time by RV1,R1 in the range 3-20 secs. The relay will thus energise for 1 sec at an interval determined by RV1.

The relay has a changeover contact which allows for connections to most wipers. The simple single speed wider is shown on Fig. 3:6a. The normally open contact on the relay is simply paralleled with the wiper switch. The 1 sec energisation of the relay thus starts the wiper, and LS1 takes it for one wipe back to park to await the next operation of the relay.

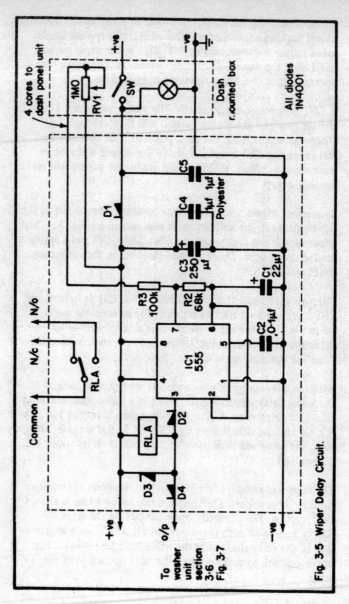

Fig. 3-5 Wiper Delay Circuit.

52

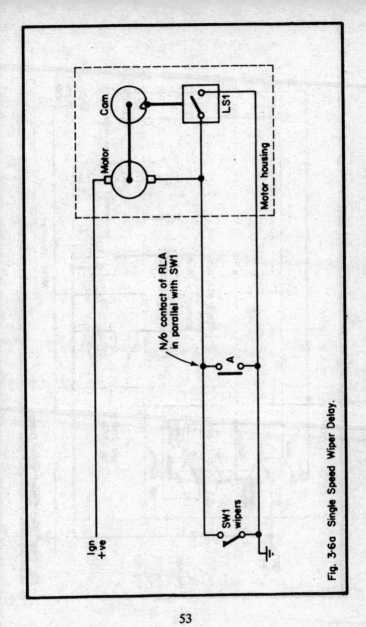

Fig. 3·6a Single Speed Wiper Delay.

53

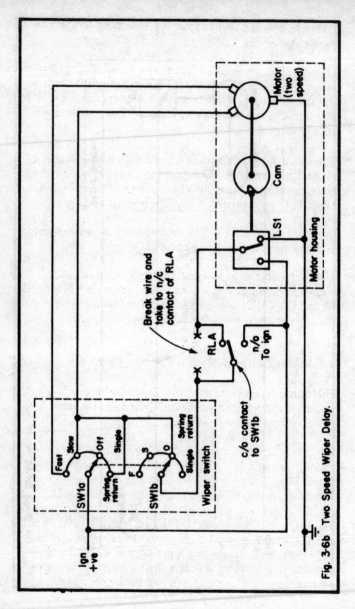

Fig. 3-6b Two Speed Wiper Delay.

54

Connection to two speed wipers is a little more complex as shown also on Fig. 3:6b.

The wire from SW1b to LS1 is cut and the relay connected as shown. As before, energisation of the relay starts the motor and the self parking LS1 carries the motor for one wipe. Note that the relay and LS1 have no effect of slow, fast or single wipe is selected.

The unit should be constructed in a diecast box with the on/off switch and delay potentiometer mounted at a convenient point on the dash panel. As RLA is only required to make, not break, the motor current, 5 amp contacts are adequate.

Diode D4 allows RLA to be energised from the washer motor as described in the next section.

The circuit obtains its supply from the ignition auxiliary supply, switched by SW. It is essential to provide a fuse. Diode D1 and C3, C4, C5 provide smoothing and protection from the nasty spikes floating around automobile supplies.

3:6. Washer Control of Wipers

This circuit is an extension of 3:5. When the washers are pressed the wipers start automatically and run for about 5 secs then stop and park.

The circuit is shown on Fig. 3:7. 1C2 is a monostable with period 5 seconds. It is triggered via D4 and the washer motor switch PB1. (If washer motor is switched on the positive side a transistor inverter TR2 as shown in the insert is used.)

The output of the monostable turns on transistor TR1, which connects to D4 on Fig. 3:5 energising the relay and starting the wipers. RV2 is an on board pre-set, and should be set so that the relay A on Fig. 3:5 de-energises half way through a wipe so that the relay does not break the motor current.

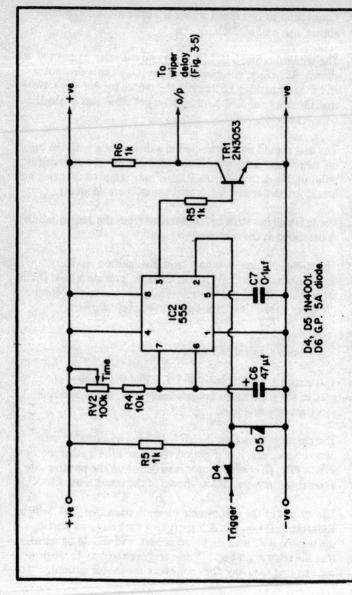

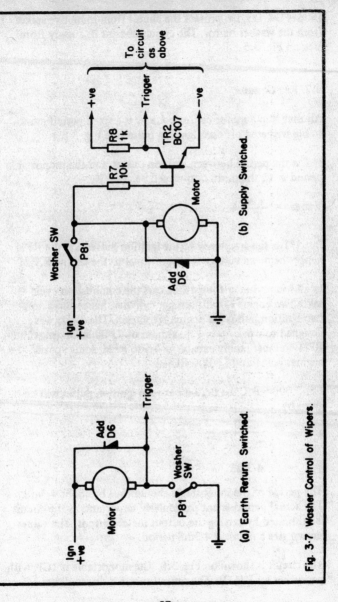

Fig. 3·7 Washer Control of Wipers.

(a) Earth Return Switched.

(b) Supply Switched.

Diodes D4, D5, D6 protect the circuit from inductive spikes from the washer motor. The circuit obtains its supply from Vcc on Fig. 3:5.

3:7. Rev Counter

All that is in a motor car rev counter is a fixed period mono-stable triggered off each ignition pulse (see Fig. 3:8.)

If T is the period between ignition pulses, and the monostable period is 't', the mean output will be given by:—

$$V\,mean = \frac{t \times V}{T}$$

But 1/T is the frequency of the ignition pulses, i.e. the RPM, hence the mean voltage is proportional to the engine R.P.M.

In all four cylinder four stroke cars the combustion cycle takes two engine revolutions per cylinder, hence there are two ignition pulses per engine revolution. The meter was designed to work up to a maximum of 7,000 RPM. (maximum RPM on most family cars being 4,500 RPM, some sports engines going up to 6,000 RPM.)

For 7,000 RPM, the period between ignition pulses will be given by:—

$$T \quad = \quad \frac{60}{2 \times 7,000} \quad secs$$

$$T \quad = \quad 4 \cdot 3mS$$

The period of the monostable should thus be about $4 \cdot 3mS$. The actual period is not particularly important, as the circuit is calibrated by setting the output meter current. The values shown give a nominal $4 \cdot 3mS$ period.

The circuit is shown on Fig. 3:8. The monostable is 1C1, with period set by R1, C1. The output meter is driven direct off the 555 output, the current being set by RV1. ZD2 defines the "high state" accurately to compensate for a varying level of high state out of the 555. It was originally designed to operate off the points contacts, sensing the opening of the points as shown in the insert, but work on the touch switch (section 5:8) showed how easy it is to trigger a 555 by induction. It was found that the circuit could operate quite happily by wrapping the trigger input around the ignition coil.

The output current is dependent on the supply voltage to a certain extent and a car battery can very between 9 and 15v. A 9v stabilised supply is therefore provided by TR1 and ZD1.

The meter can be any 0-1mA type. The traditional 270° movement meters used for rev counters will be found to be very expensive, but worth it if a good looking instrument is required. The 0—10 readings can be left, and the meter set accordingly, or the 0—10 readings replaced by 0—7 if better resolution is required.

Calibration is quite straightforward. The most accurate way is to use a signal generator set for 200Hz and set the meter for 6,000 Rpm. A simpler way is to use the car speedometer. From the car handbook you can find out what speed is equivalent to 1,000 RPM in top gear. For example the figure for the author's car is 14·6 mph per 1,000 RPM. At 60 mph the engine is thus doing 4,100 RPM. The meter could be set thus by driving at a constant 60 mph on a motorway and having a friend set the meter to the corresponding RPM. The accuracy of this method is not good, as an average car speedometer is about 5% out. The car speedometer could, of course, be calibrated first against motorway mile posts with a stop watch.

The circuit can also be used as an RPM meter for machinery by using a photo-electric or inductive pick up to sense fan blades, coupling bolts etc.,.

3:8. Car "Courtesy" Courtesy Light.

The courtesy light in a car has an annoying shortcoming, the

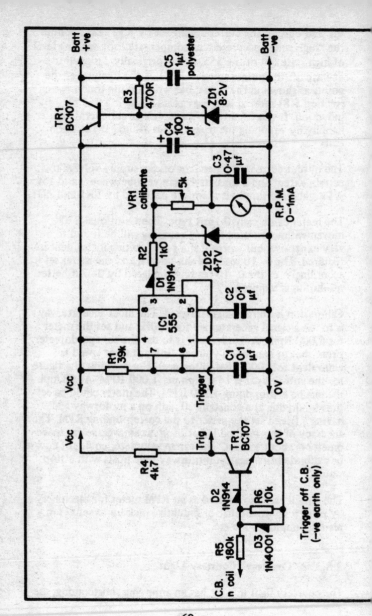

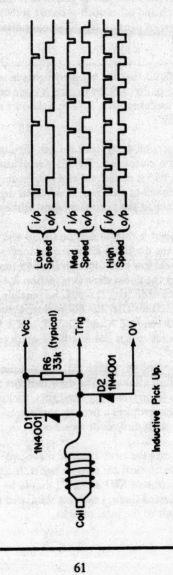

Fig. 3-8 Car Rev Counter.

Coil

Inductive Pick Up.

D1
1N4001

R6
33k (typical)

D2
1N4001

Vcc

Trig

0V

Low
Speed { i/p
o/p

Med
Speed { i/p
o/p

High
Speed { i/p
o/p

light comes on as you open the door, but when you shut the door the light goes out, just as you need it most to find the ignition switch and other such pre-start activities. This circuit keeps the interior light on for 30 seconds after the door has been opened.

The simple circuit that the courtesy lights in most cars operate with is shown in Fig. 3:9. Switching is done on the 0v side by the courtesy switches SW1, SW2 in the doors or by the interior light switch SW3.

To give a 30 second delay, lead A from the door switches is removed, and connected to the 555 circuit also shown on Fig. 3:9. The 555 is connected as a monostable, and is thus triggered by the door switches. The output drives TR1 which is connected across the normal interior light switch.

The interior light is thus turned on for 30 secs after the door is opened. If the doors are left open for longer than 30 seconds, the input is held low and although the 555 times out it will not reset until the doors close (see section 2:1) The lights will then go out immediately. It would be possible to make the lights go out 30 secs after the doors close with a single transistor between lead A and the 555, and A.C. trigger the 555, but in practice it is not worth the extra complexity.

The circuit has one advantage that was not designed in, but became apparent after use. The door switches are very prone to failure due to dirt, causing dim lights, flicking lights etc. The circuit only requires a fleeting contact, so it will work quite happily with dirty door switches.

With a bit of care the circuit can be made small enough to fit inside the base of most car interior lights. If this cannot be done, the junction of SW1 and SW2 should be located and the circuit inserted there. Lead A is then used as the output from the circuit to the interior light.

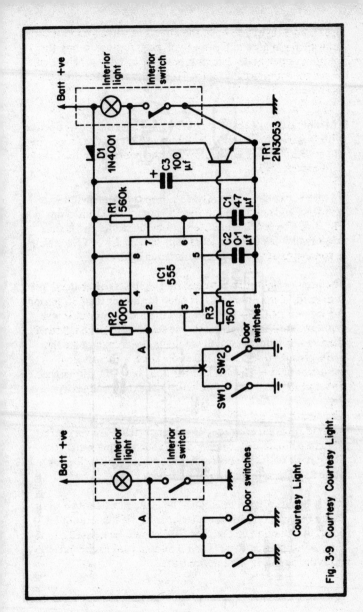

Fig. 3-9 Courtesy Courtesy Light.

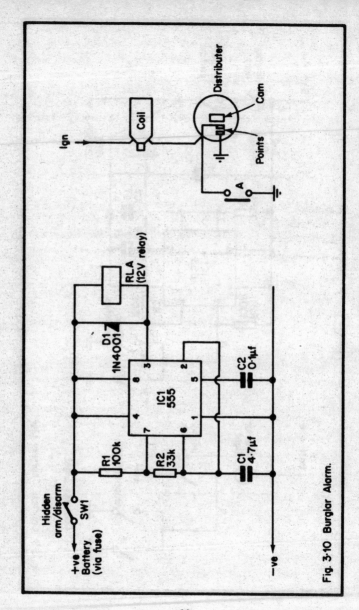

Fig. 3-10 Burglar Alarm.

3:9. Car Burglar Alarms

There are many, many schemes for car burglar alarms, one could no doubt fill a book with them. The 555 can be used in many schemes, a few examples of which are given. The field is endless, though, and the reader should be able to make an alarm as simple or as devious as he needs. Remember, though, that no alarm will beat a professional thief, these alarms are merely to deter the joyrider.

The first circuit is not really an alarm but a deterrent. It works by making the car run very rough, the resulting stalling and backfires being enough to deter any car thief. The circuit is shown on Fig. 3:10. IC1 is a 555 connected as an astable driving relay A. The timing components are chosen such that A is energised for about 0·1 sec every 0·5 sec.

The contacts of A are connected across the points in the distributor, care being taken to suitably disguise the wires with dirt. As the contacts close they cause odd timed sparks, and missed sparks, the result of which is a very sick sounding engine.

The circuit could be driven, via an enable switch, off the ignition, but this would not cover the case where a thief pulls the ignition wire off the coil and bridges the coil to the battery direct. The most foolproof (or thiefproof) method is to connect the circuit direct to the battery via a fuse and an enable switch. The current drawn is minimal (about 50mA mean) and there is no danger of flattening the battery. Regardless of how a thief bypasses the ignition, the circuit will still work. The beauty of this circuit is that the thief does not even suspect a burglar alarm. It will not, however, deter a car radio thief.

A second, more conventional, alarm is shown on Fig. 3:11. Like most alarms this is triggered off the door contacts for the courtesy light. When the door opens, relay A makes and latches, applying power to the 555. This is connected as an

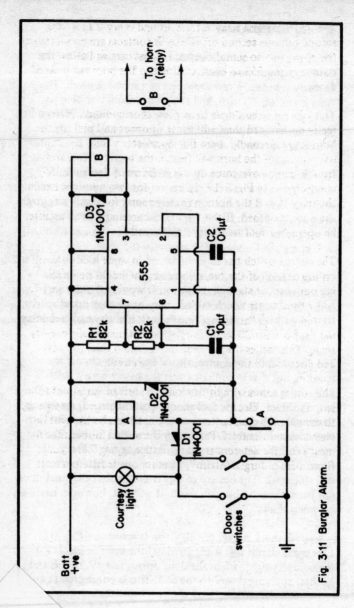

Fig. 3·11 Burglar Alarm.

astable to energise relay B for 1 second every 2 secs (one second on, one second off) Relay B contacts are connected to the horn relay to sound the horn (some cars will allow the output of the 555 to connect direct to the horn relay via a diode.)

This type of circuit does have a few shortcomings. First, a really determined thief will lift the bonnet and pull off the horn leads. Secondly, even if it does deter a thief the continual running of the horn will flatten the battery. This second trouble can be overcome by the addition of a second 555 connected as in Fig. 3:12. 1C2 is set for two minutes, hence the relay B, and the horn, will only sound for 2 minutes after the door is opened. If the thief is still around 2 minutes after he opens the door he is very determined.

The enable switch has to be mounted in some hidden position on the outside of the car, otherwise you would be unable to get out without starting the alarm. It would be quite easy to add a third timer which ran for, say, 30 seconds after being started with an "arm/reset" push button in the car, and delaying the horn for 30 seconds by another timer to allow entry again. This makes life easy for the car radio thief, however, and does add to the complexity of the circuit.

The reader can no doubt think up all sorts of variations to stop the thief. Electric fuel pump, ignition circuit, emergency flashers, headlights, battery switches, all can be brought into play to deter the thief. Personally the author thinks that the most effective deterent is a large notice saying "This car is fitted with a Burglar Alarm", whether one is fitted or not!

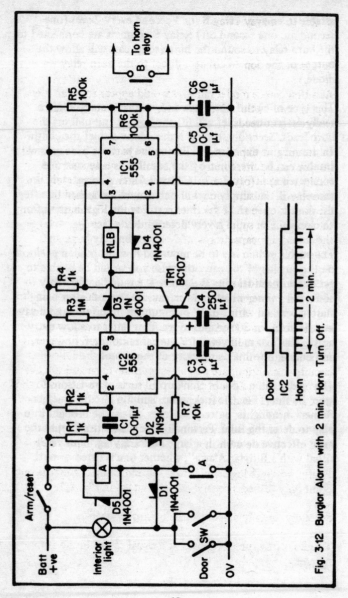

Fig. 3-12 Burglar Alarm with 2 min. Horn Turn Off.

4. Model Railway Circuits.

4:1. Introduction

At a first glance model railways would appear to be an ideal application for the 555, a 12v rail, things to control, all an experimenter needs.

In the authors experience model railways cause problems with integrated circuits. Firstly the 12v rail is usually a simple rectifier bridge (often even half wave) and is completely un-smoothed. The supply could of course, be smoothed but care would have to be taken to separate the logic supply from the locomotive supply by a blocking diode (see Fig. 4:1.) or the smoothing capacitor would be inordinately large and expensive.

Secondly, a model railway locomotive is one of the best electrical noise generators in existance. The inductance of the motor coupled with a dirty commutator, indifferent pick ups etc., leads to a very spiky current. This reflects as voltage spikes on both supply and 0V lines that can be very difficult to remove. For this reason, in the majority of the following circuits connections are made between the logic and the rail-way by isolating devices such as relays and opto-isolators, and the logic is driven off its very own supply.

For brave people who wish to mix logic and loco supplies (and it is feasible with the 555) there are a few basic rules:—

Firstly, separate logic and power 0v, and connect them at one point only (called a central earth point) as in Fig. 4:1.

Secondly, heavily de-couple the control voltage.

Thirdly, avoid running power and signal cables in the same harness.

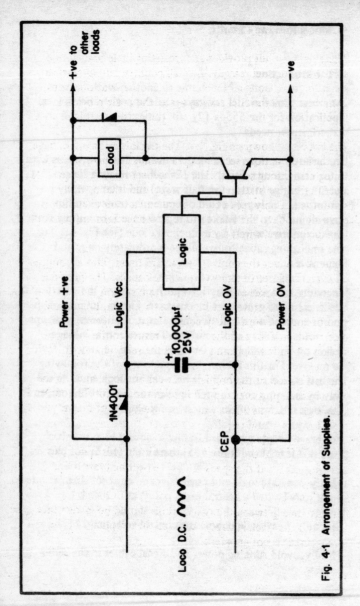

Fig. 4-1 Arrangement of Supplies.

4:2. Model Railway Shuttle Service

This simple circuit provides an interesting little branch line service for a model railway. A small country railbus starts at a station, runs along a branch line to another station, stops, then returns to the first station again, the cycle repeating indefinitely.

The circuit is shown on fig. 4:2. The track is arranged to have two isolated stations sections at each end. The power is fed to the centre long section via a changeover relay A. Diodes D1 and D2 feed the station sections and ensure that a train in station A can only move to the right and a train in station B can only move to the left. The diode connections are correct for conventional wired trains.

Relay A is under the control of the 555 timer. This is connected as an oscillator of almost equal mark space. The period is longer than it takes a train to travel from station A to station B. The train will travel until it reaches a station, then it will stop because of the diode. When the relay changes over, the diode will conduct and the train will return to the other station where it will again stop until the relay changes.

The half period of the oscillator should be made equal to the journey time plus the stop required in the station. The values shown give about 12 secs which is of the order of magnitude most layouts would require.

The stop/start of the train is unramped, but this is not particularly noticeable at the speed all self-respecting branch line trains travel.

The distance between the two stations should be made quite large, and the track suitably scenic so the simplicity of the arrangement is not apparent.

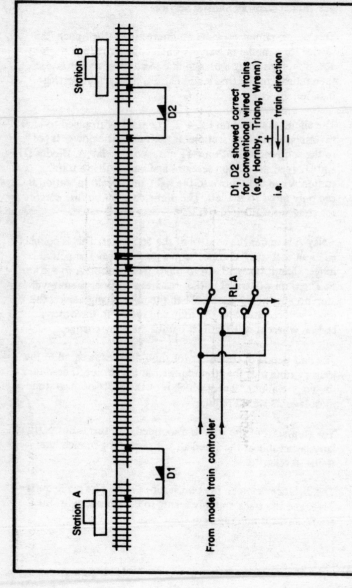

Station B

D2

D1, D2 showed correct
for conventional wired trains
(e.g. Hornby, Triang, Wrenn)

i.e. ⇅ train direction

RLA

Station A

D1

From model train controller

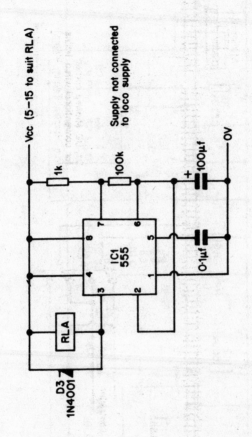

Vcc (5–15 to suit RLA)

Supply not connected
to loco supply

1k

100k

100µf

0V

7

8

6

5

IC1
555

4

3

1

2

0·1µf

RLA

D3
1N4001

Fig. 4·2 Model Railway Shuttle.

73

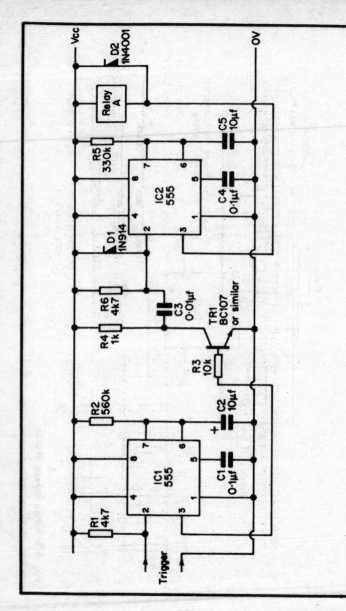

74

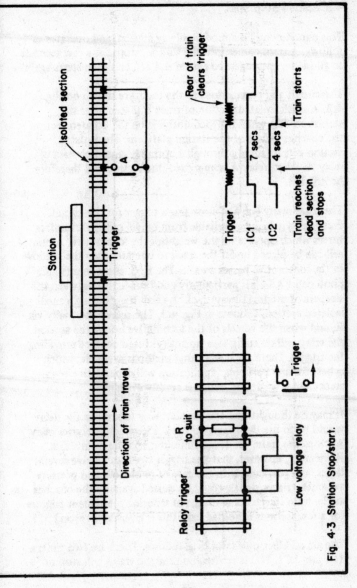

Fig. 4·3 Station Stop/start.

4:3. Station Stop/Start

This circuit controls a model railway station, and operates as
follows. A train comes into the station, stops for a few seconds
to allow the passengers to get on and off, and then starts again.

The circuit and connections to the track are shown on Fig.
4:3. A small isolated section of track is placed after the
station so that when the locomotive is on the isolated section
the coaches will be at the station platform. The isolated
section gets its supply through a normally closed contact of
relay A. When relay A is energised, the section will therefore
be isolated.

There are many ways of arranging a trigger for the circuit.
Pressure switches are available from model railway manufacturers which operate on the weight of the locomotive. A photocell can be placed under the track to operate when the shadow
of the locomotive passes over it. The R. S. components 1.C.
photocell LAS 15 is particularly good for fitting between the
sleepers of model railway track. A relay trigger using a small
isolated section is shown in Fig. 4:3. The relay is normally via
R, and when the wheels of the locomotive bridge the section
the relay will de-energise, a normally closed contact providing
the trigger. There are many other methods available, which
is best for any particular application will depend on circumstances and, no doubt, which is readily available.

It may be thought that the simplest way to arrange the delay
would be to use the trigger to fire a 555 which energises relay
A, stops the train, the timer times out and the train starts.
Remember, however, that the trigger signal can arrive several
times, and particularly in the case of photocells and pressure
switches, trigger signals could be caused again by the coaches
as the train starts. Some means of ignoring subsequent triggers
for a few seconds is needed, and this is given by a second 555.

The actual electronics works as follows. There are two timers,
1C1 and 1C2. 1C2 is set for the time the trains will stop in

the station by R5, C5 (with the values shown 4 secs.) 1C1 is set for a few seconds longer. (R2, C2 with the values shown about 7 secs.) The first trigger pulse starts 1C1. The output of 1C1 turns TR1 on, and this in turn starts 1C2. The output of 1C2 energises relay A, stopping the train. Note that 1C1 will ignore further trigger pulses for 7 seconds, after which the train will have safely left the station.

After 4 secs IC2 times out, relay A de-energises and the train starts. About 3 seconds later, IC1 times out and the circuit can be triggered again. The operation is summarised on the timing diagram on Fig. 4:3.

The circuit can be made to work with trains approaching from either direction with the track connections on Fig 4:4. The diodes bridge the relay contacts for the isolated section on the approach side of the platform according to the track polarity (and hence trains direction). The diode connections are correct for conventional wired locomotives (eg. Hornby, Triang, Atlas, etc.)

The circuit can be disabled by bridging relay A contacts or simply turning off the logic supply.

4:4. Model Railway Controller

Model railway controllers tend to be one of three types. The first, and crudest, is to have a constant D.C. voltage and to feed the locomotive through a variable resistance which controls the speed.

The second is the variable transformer. The speed control selects transformer taps which are rectified and fed to the locomotive.

The third type is similar to a variable voltage power supply, the locomotive being fed from a variable D.C. supply, usually some form of series regulator.

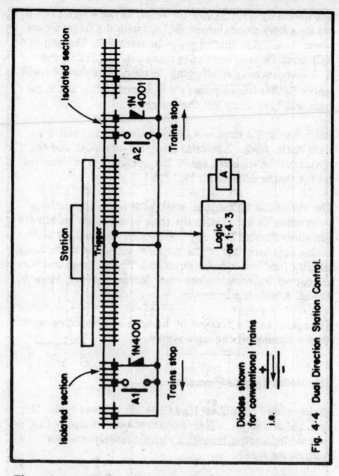

Fig. 4·4 Dual Direction Station Control.

The unit to be described works on a different principle. The locomotive supply is a constant height pulse at fixed clock rate, and the speed control varies the pulse width (see Fig. 4:5.) This form of control gives excellent slow running characteristics, a loaded goods train can run at a realistic crawl without stalling.

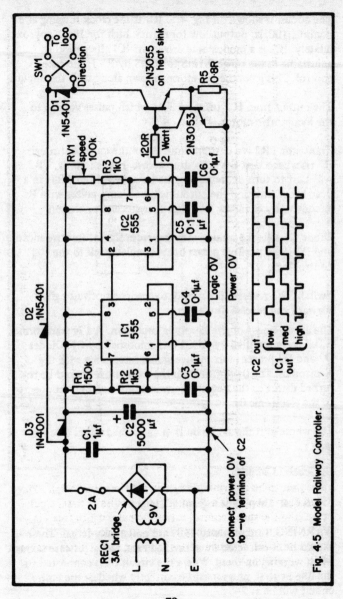

Fig. 4-5 Model Railway Controller.

The circuit is shown on Fig. 4:5. IC1 is the clock running at a nominal 100Hz, output low for 0·1mS, high for 10mS approximately. IC2 is a monostable driven off IC1. Its period is adjustable in the range 0·1mS to 10mS by RV1, the speed control. This gives the waveforms shown along with the circuit.

The output from IC2 turns on TR1 which pulses voltage to the locomotive motor.

Transistor TR2 is a current limit. When the current through R5 rises such that the volts drop across it rises to 0·8v, TR2 will start to turn on, removing the base drive from TR1. In a dead short situation the output will be 1 amp pulses with R5 as shown. By changing R5 other currents can be chosen.

Diode D1 kills the inductive spikes from the locomotive motor, and D2 stops positive pulses being coupled back to the control logic.

Switch SW1 reverses the supply to the locomotive to give forward and reverse directions.

The arrangement of the supply is important. A 12v transformer is full wave rectified by REC1 and smoothed by C1. Diodes D2 and D3 prevent locomotive spikes interfering with the electronics. The 0v return from the locomotive should be returned direct to the negative terminal of C1 and not connected to the electronic 0v.

The whole unit should be built in an earthed metal case for safety.

5. General Circuits

5:1. S.C.R. Drive (Photographic Timer)

WARNING. Circuits using 240v are potentially lethal. These should be constructed in an earthed case, and great care taken when working on them. Where the circuit 0v is connected to the line neutral, plugs must be correctly wired or the entire circuit will be at 240v.

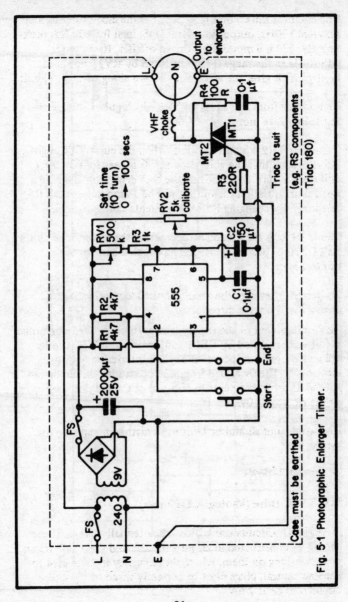

Fig. 5-1 Photographic Enlarger Timer.

81

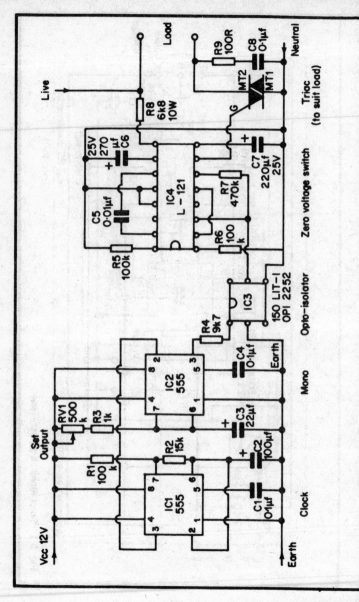

82

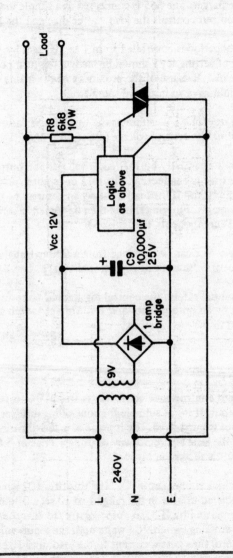

Fig. 5·2 Burst Fired Heater Control.

This simple circuit shown in Fig. 5:1 gives control of a photographic enlarger. The 555 is connected as a simple monostable, but the output controls the load via the diac and the triac.

The delay period is controlled from 1 to 100 secs by RV1. For ease of setting RV1 should be made a ten turn pot with analogue dial. Because of the proximity of the mains voltage RV1 should have an insulated spindle.

Variable resistor RV2 calibrates the timer for variations in Ct by adjusting the control voltage (see section 2:9)

The triac is effectively burst fired. When the 555 output is high, the triac is conducting. The turn on is not synchronised to the supply and if the load is, heavy spikes may be induced onto the mains. Filtering should then be placed in series with load (R.S. components T.V. chokes)

The supply Vcc can be obtained from a simple transformer/ bridge/rectifier/smoothing capacitor circuit.

For maximum safety, the control logic could be connected to the triac by an opto isolator and zero voltage switch as described in the next section.

5:2. Heater Control

This circuit controls an electric heater from 10% to 100% of rated output. It is not suitable for controlling drills or lights as the output is burst fired. Use is made of a zero crossing switch, which is the least antisocial way of using a Triac or S.C.R. The circuit is shown on Fig. 5:2.

IC1 is a clock running at 0·1Hz, and this fires IC2 which is a monostable adjustable in the range 1 to 10 secs. The output of IC2 is connected by IC3, an opto isolator, to energise, IC4, the zero crossing switch. IC4 waits until the mains voltage crosses zero, then passes current to the load until IC2 times out. Because IC4 delays the turn on until zero voltage there are no transients induced on the mains.

The load is thus energised for between 1 and 10 seconds every 10 secs giving control over the heat output.

The use of the opto isolator IC3 separates the control logic from the mains for safety.

5:3. Motor Underspeed Alarm

The re-triggerable monostable described in section 2:6 can be used to give warning of motor underspeed. The circuit is shown in Fig. 5:3.

Shaft pulses are obtained from some suitable transducer, such as a vane switch or photocell. These re-trigger the 555, discharging the timing capacitor as described in section 2:6 so the output stays permanently high.

Should the motor speed fall, the pulses will not be fast enough to discharge the timing capacitor, and the 555 will give pulses out.

These pulses can be used to drop out a latching relay or set an alarm flip flop as shown.

The circuit is also used in computer control applications to indicate that the computer is healthy. Every 0·1 secs., (or some time to suit) the computer sends a signal which re-triggers the monostable. Failure of the computer means the monostable is not re-triggered, so it times out giving indications of computer failure. This circuit is called a Watchdog Timer.

5:4. Overspeed warning

Warning of motor overspeed can be given by using two timers in the re-triggerable mode, as shown on Fig. 5:4.

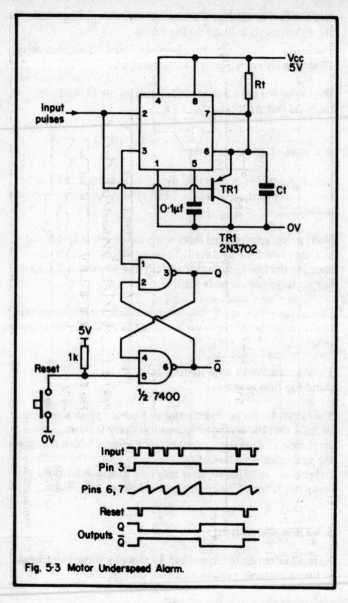

Fig. 5-3 Motor Underspeed Alarm.

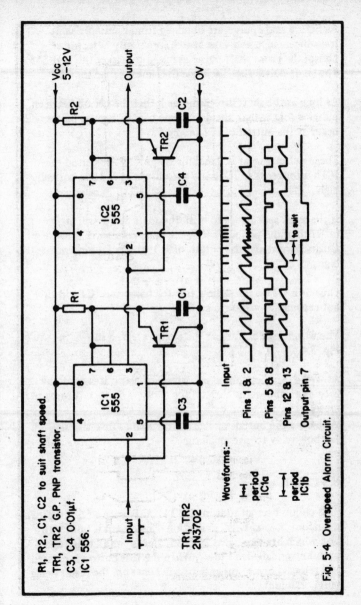

R1, R2, C1, C2 to suit shaft speed.
TR1, TR2 G.P. PNP transistor.
C3, C4 0·01μf.
IC1 556.

TR1, TR2
2N3702

Waveforms:-

period IC1a

period IC1b

Input

Pins 1 & 2

Pins 5 & 8

Pins 12 & 13

Output pin 7

to suit

Fig. 5·4 Overspeed Alarm Circuit.

As before shaft pulses are obtained from a suitable shaft transducer, such as a vane switch, proximity detector or photocell. These shaft pulses are used to trigger the first 555. The shaft pulses also discharge the timing capacitor C1.

As long as the shaft is rotating such that the period between pulses is longer than the delay time of IC1, regular pulses will occur at the output of IC1 on pin 5.

These pulses trigger IC2 and discharge C2. The period of IC1b is longer than IC1a and the output of IC1b is normally high.

If the shaft speed goes to high, the shaft pulses will arrive faster than the period of IC1, the shaft pulses will then keep C1 discharged and the output of IC1 will lock up permanently high.

Capacitor C2 on IC2 is now free to charge and IC2 will time out causing its output to go low.

The waveforms existing in the circuit are also shown in Fig. 5:4.

As well as applications as an overspeed alarm it can also be used as an automatic turn on for teletype motors, turning the motor off again when data transmission ends.
As before the output of IC2 can be used to de-energise a latching relay to give an alarm.

5:5 Coin Tosser

This circuit uses an additional TTL I.C.P. with a 555 to provide a random "heads/tails" indication. The circuit is built into a box which has two lights labelled "heads" and "tails" and a push button labelled "toss". When the toss button is pressed and released one or other light will come on, the result being purely random.

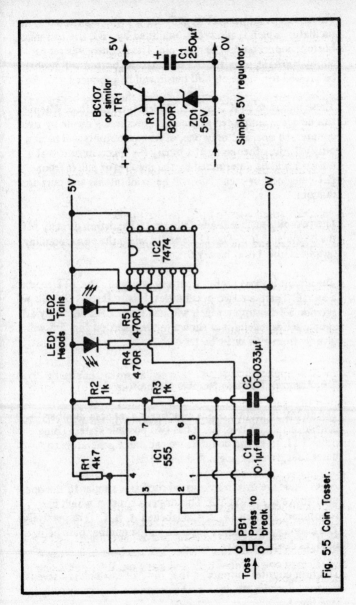

Fig. 5-5 Coin Tosser.

89

The circuit is shown on Fig. 5:5. IC1 is a 555 connected as an oscillator, which is allowed or inhibited by PB1, the toss push button, connected to the reset pin. The frequency is set for about 100KHz, so in the 0·5 sec. that the button will probably be pressed for, some 50,000 pulses will be produced.

These pulses go to IC2, a TTL 7474 D type flip flop. When this device is connected as shown, it becomes a divide by two counter, so one, and only one, of its two outputs will be at a binary 1, the other being at a binary 0. Which output is at a binary 1 will be determined by the number of pulses from IC1 being odd or even. This will be to all intents and purposes random.

The two outputs heads and tails are LEDs, driven directly off the outputs of IC2. The LEDs are lit when the corresponding output is low, i.e. at binary 0.

The circuit is most useful if battery powered. The TTL requires a 5v rail. There are two possibilities. Firstly three 1·5v cells will provide a 4·5v supply which will just work TTL. Secondly a simple series regulator as shown in the insert on Fig. 5:5 will give 5v from a 6v or 9v battery.

5:6. Assorted Random Number Generators

The circuit in section 5:5 is effectively a random number generator to base two. (ie. it has two possible states) Using other TTL gates and counters we can make generators to bases four, five, six, eight and ten.

All of these are driven by a gated oscillator similar to the one used above in section 5:5. This drives a counter which has four binary outputs (1, 2, 4, 8 labelled A, B, C, D respectively). The counter, and wiring, depends on the number base being used, as described later.

The four counter outputs go to a 7447 I.C., which is a seven segment decoder, to drive a seven segment display directly. (see Fig. 5:6;)

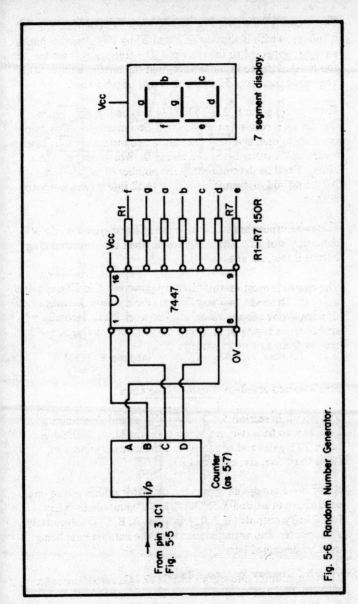

Fig. 5-6 Random Number Generator.

7 segment display.

R1–R7 150R

7447

Counter
(as 5-7)

From pin 3 IC1
Fig. 5-5

91

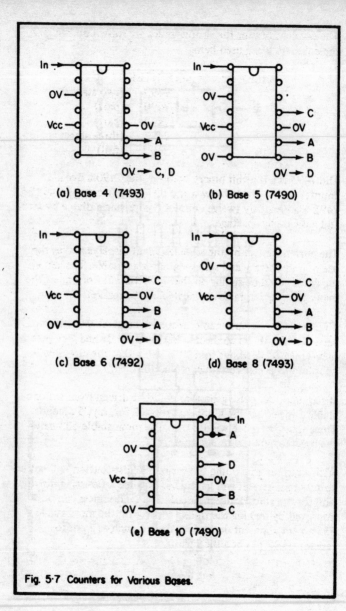

(a) Base 4 (7493)

(b) Base 5 (7490)

(c) Base 6 (7492)

(d) Base 8 (7493)

(e) Base 10 (7490)

Fig. 5·7 Counters for Various Bases.

92

The connections for the various bases are shown on Fig. 5:7, the counters being used being:—

BASE	COUNTER	
4	7493	(two stages)
5	7490	(part)
6	7492	(part)
8	7493	(three stages)
10	7490	(full)

The 7493 is a four bit binary counter, the 7490 a decade counter with a divide by two and divide by five stage, and the 7492 is a divide by twelve counter comprising a divide by two and a divide by six stage.

The current drawn by the seven segment displays makes the use of a 9v battery and series regulator advisable. The series regulator could be similar to that in section 5:5 or one of the many I.C. regulators now available on the market.

TTL counters are vulnerable to supply noise. A $0.01uF$ capacitor should be connected between the 0v and Vcc pins on each I.C.P.. Without these it is likely that the circuit will be found to prefer certain combinations.

If required, the pulse generator could be driven from a monostable 555 so that the numbers "rotate" for, say, 5 seconds. The circuit in fig 5:8 will do this. The monostable 555 gates the oscillator on the reset as before.

With a monostable doing the gating the distribution will not be strictly random. (It will have what is called a Gaussian distribution for the statistically minded!) The randomness can be improved by making the timing resistor on the monostable 555 an environment dependent device, such as a photo resistor (ORP 12) or a thermistor.

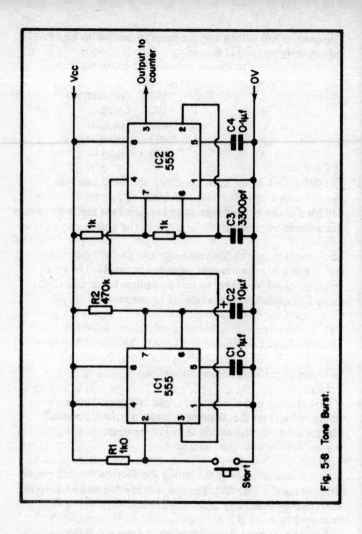

Fig. 5·8 Tone Burst.

5:7. Electronic Dice

One special random number generator is an electronic dice.
This is a box with seven L.E.D.s arranged in the pattern shown
on Fig. 5:9, and a "roll" push button.

94

The random number generator is a 555 oscillator and a divide by six counter (7492) as in section 5:6. If we look at the pattern required for the six states of the lamps we find;

Binary CBA	State	Lamps On
001	1	a
010	2	bc
011	3	a,bc
100	4	bc,de
101	5	a,bc,de
000	6	bc,de,fg

From this we see that lights b and c, d and e, f and g are always on as pairs. Comparing the lamps with the binary states we see that:—

Lamp a	corresponds to output A
Lamps bc	corresponds to output B or C or "6"
Lamps de	corresponds to output C or "6"
Lamps fg	corresponds to "6"

Where "6" is given by $A = B = C = 0$

The logic to do this is provided by two additional I.C.s connected as in Fig. 5:9. Both are TTL logic ICs, IC3 being a 7404 inverter and IC4 a 7410 triple 3i/p gate. The seven output lights are L.E.D.s driven direct off the TTL gates, hence a logical 0 is required to light the lamp.

IC4a gives '0' out for $A = B = C = 0$ (i.e. "6"). IC4b gives '1' out for B or $C = 1$ or IC4a output a '0' (i.e. "6"). IC4c gives a '1' out for $C = 1$ or "6". The outputs of IC4b and c have to be inverted by IC3d, e to drive the L.E.D.s.

The circuit is built into a box with the seven L.E.D.s arranged in the usual dice pattern and the roll switch mounted on the top.

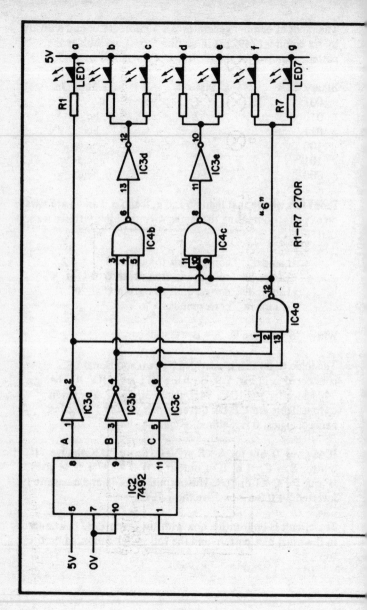

R1–R7 270R

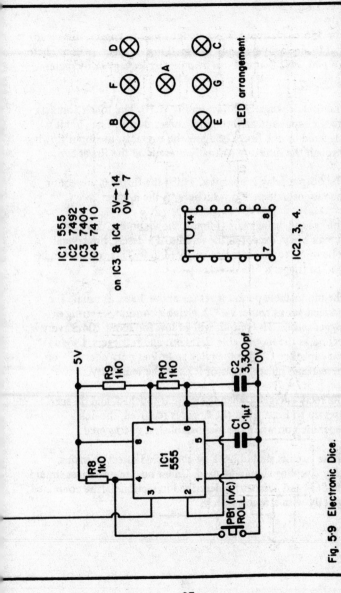

LED arrangement.

```
D ⊗        ⊗ C
F ⊗    ⊗ A    ⊗ G
B ⊗        ⊗ E
```

IC1 555
IC2 7492
IC3 7404
IC4 7410

on IC3 & IC4 5V → 14
              0V → 7

IC2, 3, 4.

Fig. 5-9 Electronic Dice.

97

5:8. Touch Switch

The 555 needs about 0·5uA at the input to trigger, this means that it can easily be made to function as a touch switch. There are two ways that this can be done, by leakage or by mains hum.

The first, leakage, is shown on Fig. 5:10. The touch contacts are the input and an 0v connection as shown. The circuit is triggered by the finger bridging the contacts, the input flowing through the moisture naturally present on the finger tips.

The output relay is energised whilst the finger is present or the monostable period, whichever is the longer.

The second, using mains hum, is shown on Fig. 5:11. The human body, except in the middle of a desert, has several volts of 50 to 60 Hz mains enclosed in it. This voltage can be used to trigger a 555 direct.

The monostable period is set for about 1 sec. as usual. The induced mains comes via C2, giving a continuous string of trigger pulses. The output will go low for about 10mS every second as the monostable times out and retriggers. Diode D1 and capacitor C3 slug the relay so it does not "chatter", on these 10mS pulses. Resistor R2 sets the sensitivity.

The relay energises when the plate is touched, and de-energises up to 1 sec. after the finger is removed, the delay depending on when the monostable last re-triggered.

These circuits should NOT be connected direct to mains, relay coupling should be used. Under no circumstances should the SCR and Triac drive described in section 5:1 be combined directly with a touch switch.

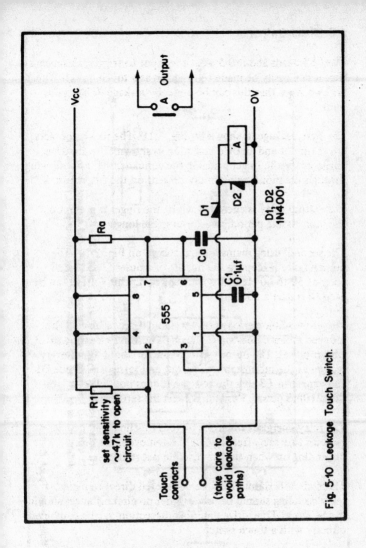

Fig. 5-10 Leakage Touch Switch.

99

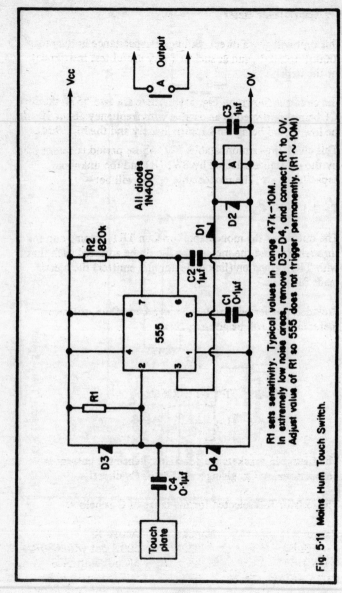

R1 sets sensitivity. Typical values in range 47k–10M.
In extremely low noise areas, remove D3–D4, and connect R1 to 0V.
Adjust value of R1 so 555 does not trigger permanently. (R1 ~ 10M)

Fig. 5·11 Mains Hum Touch Switch.

5:9. Capacitance Meter

This unit will give a direct reading of capacitance in the range
1000uF to 10uF, and as such is a very useful test instrument
for the bench.

The circuit is shown on Fig. 5:12. There are two 555s, the first
IC1 being connected as an astable with frequency about 100Hz,
the low period being 1mS approximately and the high 9mS.

This clock fires a monostable, IC2, whose period is determined
by the resistance selected by SW1 (R) and the unknown
capacitance Cx. The monostable period will be:—

$$T = 1.1 \times R \times Cx.$$

The output of the monostable turns on TR1, passing current
through Rm and the meter. Rm should be adjusted such that
with TR1 turned on (link collector and emitter) the meter
reads full scale.

Provided the period T is in the range 1 to 10mS, the mean
meter current will be given by:—

$$I = \frac{T}{10} \quad mA$$

$$\text{but} \quad T = 1.1 \times R \times Cx$$

$$\text{hence} \quad I = \left[\frac{1.1 \times R}{10} \right] \times Cx$$

The figures in brackets are constants, hence the current is
proportional to Cx, giving a reading of Cx directly.

The resistor R is selected for the range of C as below:—

Range	Nominal R	Actual R
1 − 10uF	820R	500R pot with 680R
0.1 − 1uF	8K2	5K pot with 6K8
0.01 − 0.1uF	82K	50K pot with 68K
1000pF − 0.01uF	820K	500K pot with 680K

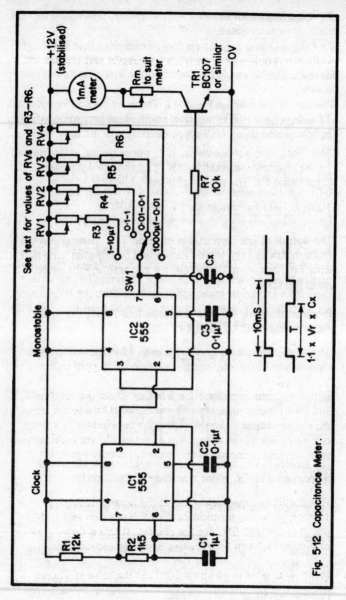

Fig. 5-12 Capacitance Meter.

102

By using a trim pot in series with a resistor, calibration for each range is obtained.

Calibration of the meter is simple. Collector and emitter on TR1 are linked, and Rm adjusted so the meter reads 10 (full scale).

Trim pots RV1 – RV4 are now adjusted on known capacitors as below:–

1 – 10uF range	use	4·7uF Tantalum (± 20%)
0·1 – 1uF range	use	0·47uF Polyester (±10%)
0·01 – 0·1uF range	use	0·047uF Polyester (±10%)
1000pF – 0·01uF range use		4700pF S.M. or
		Polyestyrene (±1%)

In each range, the corresponding trimpot is adjusted to give 4·7 on the meter. The capacitors chosen above give reasonable accuracy at low cost. For maximum accuracy the calibration capacitors should be measured elsewhere then the trim pots set accordingly. Electrolytics should not be used for calibration because of the wide tolerance.

The supply should be a well regulated 12 volts, as the meter current will be directly proportional to the supply volts.

In use, an unknown capacitor is placed across the terminals and SW1 rotated until a reading obtained (Outside the range the meter will read 1 or 10 depending on whether it is over or under). With high voltage electrolytics a lower reading than nominal is normal as the capacitor is only operating at about 6 volts.

It is possible to read capacitors in the range of 100pF to 1000pF by having an additional switch position and a 1MΩ pot in series with 8M2 for the resistor. If this is done, great care must be taken with leakage and stray capacitance.

It is not possible to go above 10uF, as the value of R goes

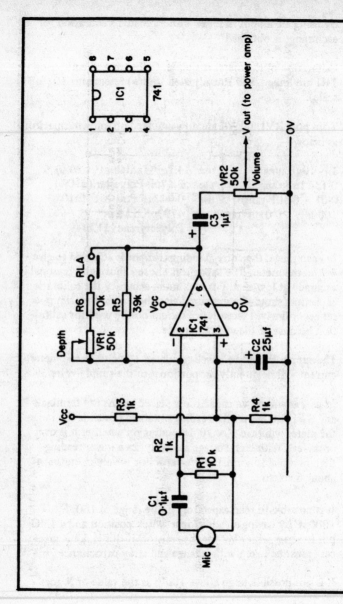

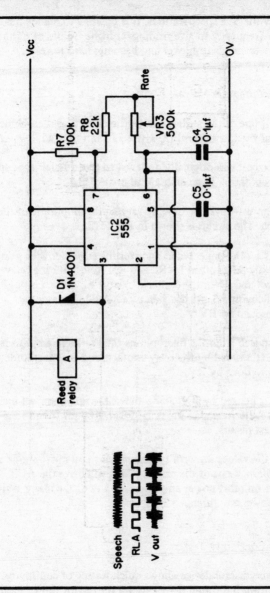

Fig. 5·13 Computer Voice.

105

below the 555 specifications. It is possible to reduce the clock frequency to give readings at higher values of C, but the meter tends to flicker and becomes hard to read.

5:10. Computer Voice

This is one of the author's favourite circuits as it combines two of the most useful I.C.s, the 555 and the 741.

The circuit was originally designed to give a Dalek type speech for a computer in an amateur dramatics play.

The circuit works by quickly varying the amplitude of the speech. The circuit is shown in Fig. 5:13.

IC1 is a 741 connected as an inverting amplifier with gain normally determined by R2 and R5. Across R5 is a relay contact and when the contacts close RV1 and R6 is put in parallel with R5 and the gain is reduced by an amount determined by RV1.

The speech from the microphone thus varies in amplitude at a rate determined by the relay contacts and the magnitude of the variations by RV1.

RLA is buzzed by IC2, a 555 timer, the rate being set by RV3. A frequency in the range 20–60Hz was found to give the best results.

With the values shown a typical moving coil microphone will give 500mV out of the circuit which will drive the AUX inputs on most power amplifiers, or any of the many available I.C. power amplifiers.

5:11. Continuity Tester

This very useful device allows quick testing of continuity, and the author has found it very useful for ringing out of multicore cables. The device consists of a box with two external test leads labelled positive and negative. If a resistance is connected between the test leads, the internal speaker sounds giving a note as below:—

Short circuit (leads linked)	fo	2KHz
5K	fo	1KHz
80K	fo	100Hz

The current flowing from the leads is small, and the circuit can be used around semiconductors without fear of damage.

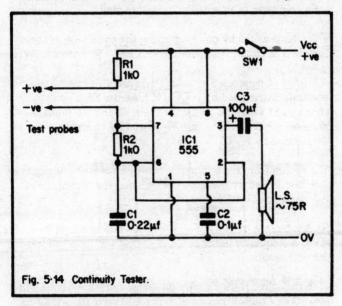

Fig. 5·14 Continuity Tester.

The circuit is shown on Fig. 5:14. It is a conventional astable 555 circuit with the test leads in series with the charging resistor, the output driving a small loudspeaker.

Because the current through the test leads is undirectional, from positive to negative, the circuit can be used for the testing of diodes or simple "diode" testing of transistor junctions.

The use of the tester for on board testing should be tempered with care. Unexpected results can be obtained with all impedance measuring devices because of sneak paths via power supplies and the like.

5:12. Signal Injector

This unit is a signal injector for use in testing audio and other amplifiers. It gives a square wave out which is rich in harmonics, the frequency of which can be varied from 50Hz to 15KHz.

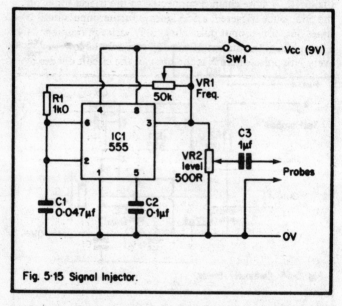

Fig. 5·15 Signal Injector.

The circuit is shown in Fig. 5:15. It is a 555 astable connected in its equal mark/space mode. The frequency is set by VR1, R1 and C1 in the above mentioned range.

VR2 controls the output level, the output being A.C. coupled by C3.

To prevent stray radiation into the circuit under test, the circuit should be constructed in a die-cast box with the output being taken via co-ax cable.

The current drain is minimal, and the unit will run for months on a 9v battery.

5:13 Frequency Divider Circuit

The 555 can be used as a frequency divider as shown on Fig. 5:16. A pulse chain is connected to the trigger input of the 555. Once triggered, a 555 ignores further inputs until it times out. If the input pulse chain only varies in frequency by a few percent the output of the 555 will only be low, at every Nth pulse where N is the number the circuit divides by.

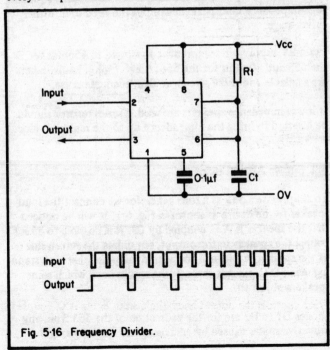

Fig. 5·16 Frequency Divider.

The period of the 555 should be (N−0.5) times the period of the input frequency, where N is the ratio of the division. For practical applications N should be kept at 5 or less.

The input frequency can vary by 50/N percent and the circuit will still remain locked. It is best suited for division of stable frequencies such as mains 50Hz or a crystal oscillator.

6. Alarms and Noise Makers.

6:1. Introduction

The maximum load current of the 555 is 0.2A and the 556 is 0.15A. This means that the devices can be coupled to small high impedence loudspeakers without the need of a further driver stage.

For a 9V supply, these currents correspond to 45ohms for the 555 and 60ohms for the 556. Many 75ohm loudspeakers are available and these give more than adequate noise.

If lower impedence speakers are used, a series resistor should be inserted to bring the impendence up to the required value.

6:2. Simple Alarm Circuit

To simply use a 555 as a tone generator we connect the loudspeaker to the device as shown in Fig. 6:1. It will be noticed that the speaker is A.C. coupled by C3. It is possible to direct couple the speaker to the output, but unless the return side of the speaker is returned to the mid rail point there is a standing mean D.C. current through the speaker coil which some speakers object to.

Diodes D1 to D3 are for the protection of the 555 from any inductive spikes caused by energising and de-energising the

110

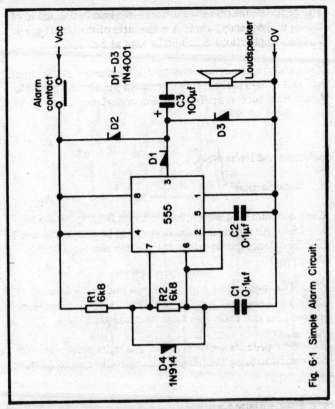

Fig. 6·1 Simple Alarm Circuit.

speaker coil. (see section 2:13) In the authors experience they are not necessary except in the case of unusually highly inductive speakers. They are omitted on all subsequent speaker driving circuits. Diode D4 is added if equal mark space is required. (see 2:12) Alternatively the equal mark space circuit described in section 2.2 can be used.

The cost of a simple 555 tone generator compares vary favourably with the commercial audible warning devices, with the added advantage that the pitch can be varied where more than one alarm is required. If two alarms cannot occur together, considerable cost saving can be made by using the circuit of

Fig. 6:2. Different timing resistors are connected to the alarm contacts and the 555 itself gets its current through D1 etc.

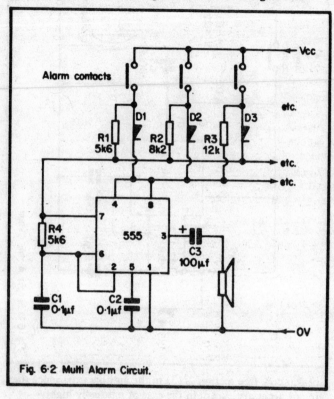

Fig. 6·2 Multi Alarm Circuit.

6:3 Bleeper Circuit

In noisy areas it is sometimes difficult to hear a steady tone. A bleeper, giving short tone bursts, is much better.

Two 555s (or one 556) are used. The first (1C1) is connected as an oscillator with a charge time (output high) of about 1 sec. and a discharge time (output low) of about 0·2 sec.

The second 555 (IC2) is connected as an oscillator with the frequency of the tone required. The output of this oscillator is connected to the loudspeaker in the usual manner.

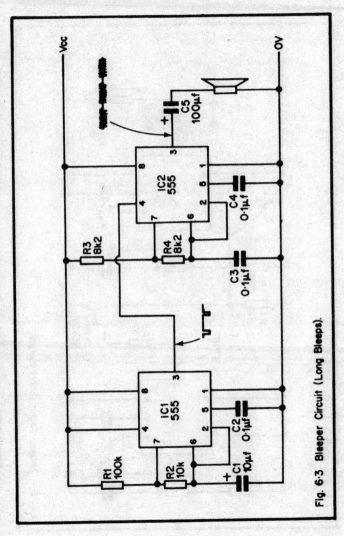

Fig. 6·3 Bleeper Circuit (Long Bleeps).

If the output required is a long burst with a short gap, the output of IC1 is connected to the reset pin of IC2, see Fig. 6:3, so whilst the output of IC1 is low the oscillations of IC2 are inhibited.

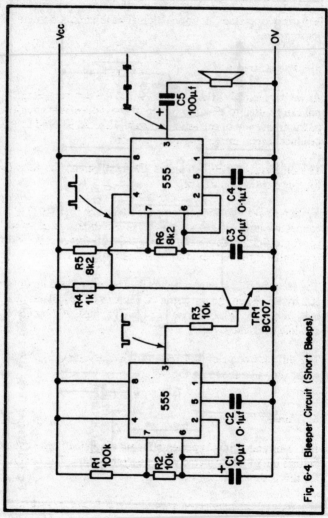

Fig. 6·4 Bleeper Circuit (Short Bleeps).

114

To give a short burst with a long gap, a transistor and two additional resistors are inserted between IC1 and IC2. (see Fig. 6:4.) This inverts the output of IC1, hence the oscillations of IC2 are inhibited during the long period when IC1 is high and allowed when IC1 output is low. If the alarm circuit is to be battery operated this connection gives minimum current consumption.

6:4. Police Siren

As we have already seen, (2:9 and 2:10) the control voltage pin can be used to alter the period of the monostable circuit or the frequency of the astable circuit. This can be used to produce several interesting alarm circuits.

The simplest is the British police "De-Dah" siren. The circuit for this is shown on Fig. 6:5.

Again we have two 555s (or one for 556). The first (IC1) is connected as an oscillator with charge and discharge times set for about 0·5 sec. Diode D1 gives equal mark/space ratio which is essential if the output is to sound right.

The second (IC2) is connected to the control voltage input of IC2 by R5. When the output of IC1 is low, IC2 oscillates at a high frequency. When the output of IC1 is high, IC2 oscillates at a low frequency.

The shift in frequency is determined by the value of R5, the lower the value the more the shift.

6:5. Siren

If a rising and falling exponential ramp is applied to the control voltage pin of an oscillator we get a siren output. The circuit for this is shown in Fig. 6:6.

IC1 and IC2 are connected as in section 6:4. The output of

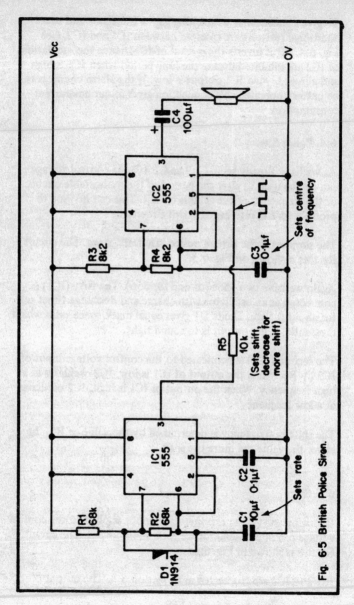

Fig. 6-5 British Police Siren.

116

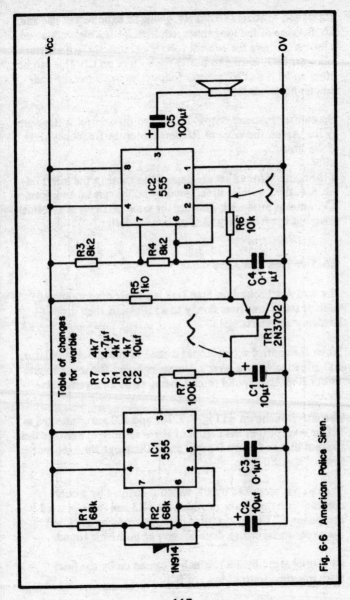

Table of changes
for warble

R7	4k7
C1	4·7μf
R1	4k7
R2	4k7
C2	10μf

Fig. 6·6 American Police Siren.

117

IC2 is used to charge C1 via R7, giving an exponential rise and fall. Because of the long times required, R1 is a high value and is unable to drive the control voltage pin directly. An emitter follower, TR1, is used to buffer the voltage on C1. This can be either an NPN or PNP emitter follower to suit what the constructer has available.

The emitter follower output is taken to the control voltage pin by R6. Again, the value of R6 determines the frequency shift of the siren.

If the component values are changed to those in the insert on Fig. 6:6, the "warbler siren" used by police cars in American T.V. series is produced. This has the same make up as a normal siren, but the frequency shifting is much faster.

6:6. Star Trek Red Alert

The Red Alert sound on Star Trek is a sawtooth modulated tone. It starts low, rises slowly to a high value then drops suddenly and starts again.

When the control voltage input is used to control an oscillator, an increasing voltage gives a decreasing tone. The input to give us our Red Alert sound should therefore be an inverted sawtooth.

The circuit is shown on Fig. 6:7. IC1 and IC2 are connected as before, except that the period of IC1 is set to give a charge time (output high) of about 0·5 sec. and a Discharge time (output low) of about 0·1 sec.

The ramp is produced by C1, which is charged by a crude constant current source provided by TR2 and its associated components. A simple resistor charging could be used, but the exponential charge does not give an authentic sound.

C1 is discharged by TR1, which is turned on by the brief discharge time (output low) of IC1.

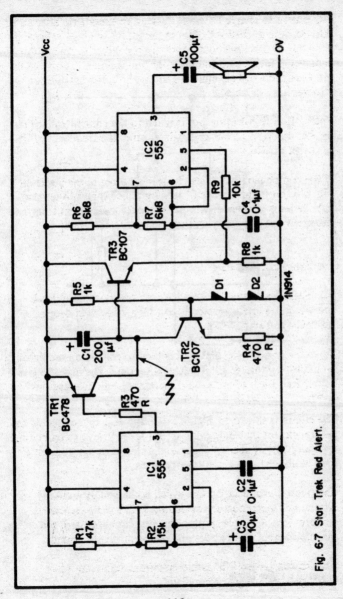

Fig. 6·7 Star Trek Red Alert.

119

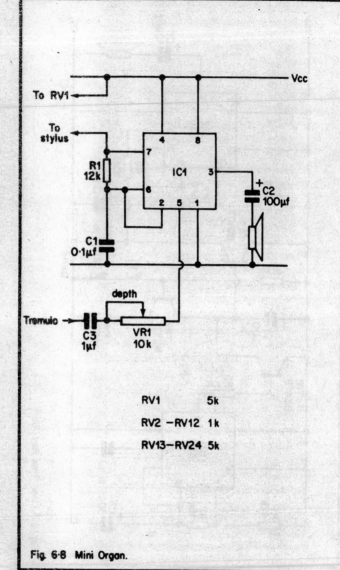

Fig. 6·8 Mini Organ.

120

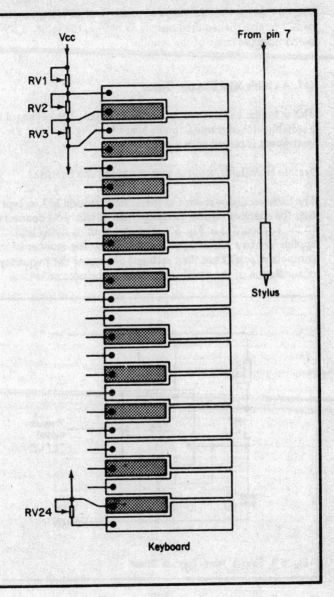

The resulting sawtooth is applied to the control voltage input of IC2 to give the required sound. As before, the value of R9 determines the range of the sound.

6:7. A Childs Mini Electric Organ

This is similar to the commercial childs organ. The keyboard is a suitably etched printed circuit board (see Fig. 6:8) and the instrument is played with a stylus.

Tremlo is available, and the amount added can be varied.

The basic oscillator is our old friend from section 6:2, except that the charging resistor is a long chain of trim pots connected to the keyboard as in Fig. 6:8. The stylus on its flying lead applies Vcc to a selected point in the chain. The number of trimpots selected (and their settings) determine the frequency of oscillation of the oscillator and hence the note produced.

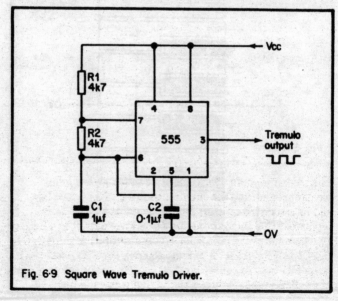

Fig. 6·9 Square Wave Tremulo Driver.

The tremulo waveform is applied to the control voltage input via VR1 causing the tone to shift up and down quickly. The setting of VR1 determines the 'depth' of the tremulo effect.

Two tremulos are possible. The first is a square wave from a 555 connected as IC1 in Fig. 6:9. This is identical in operation to the police siren circuit in section 6:4 except the shift rate is much higher.

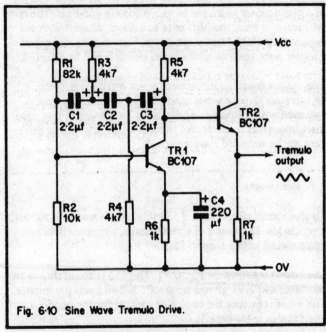

Fig. 6·10 Sine Wave Tremulo Drive.

The second tremulo uses a sine wave which gives a more pleasant sound, and the circuit is shown on Fig. 6:10. TR1 and its associated resistors are connected as a phase shift oscillator. This circuit produces a sine wave output at a frequency where the phase shift from TR1 collector back through C1, C2, C3 to TR1 collector again is 360 degrees. The phase shift from TR1 base to collector is 180 degrees, hence the frequency of oscillation is determined by the frequency at which the phase shift through C1, C2, C3 is 180 degrees. Further analysis of this circuit will be found in any elementary electronics text book.

The output of the phase shift oscillator is fed to the control voltage input via a buffer emitter follower and VR1.

The setting up of the instrument should be done with the aid of a piano unless you have an exceptionally good ear. The trimpots interact from the high note end down, so start with the highest note (RV1) and work down. Once set, do not adjust a higher note again or all the notes below it will be affected.

The output from the instrument is not equal mark/space, in fact the mark/space varies according to the note. A version was tried with a subsequent divide by two stages to give equal mark/space but it was found to make surprisingly little difference to the sound of the instrument.

6:8. Metronome

To give sound effects with a 555 it is not necessary to use high frequencies. The sound of a metronome, a simple click, can be produced with a single 555.

The circuit is shown on Fig. 6:11. The 555 is connected as an oscillator and RV1 gives a range of 10—140 beats per minute. The notes regarding the choice of speaker for the previous circuits also apply here.

If the cone of the speaker is accessible, the tone of the metronome can be made indistinguishable from a clockwork metronome by doping the cone with polyurethane varnish. Care should be taken not to allow the varnish to jam up the coil movement.

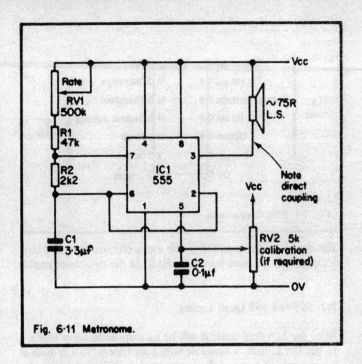

Fig. 6·11 Metronome.

7: 555 Variations

The 555 has a family of related devices, the 556, 558, 559 timers. These are basically several 555s in one case.
(N.B. 558 and 559 supercede 553 and 554 devices and are pin compatible)

7:1. 556 Timer

The 556 is a dual 555 timer in a 14 lead dual in line pack. Its connections are shown on Fig. 7:1.

Its operation is identical to the 555, all named connections fulfilling the same functions. The one difference is that it can only source or sink 150mA (0·15A) compared to the 200mA of the 555.

125

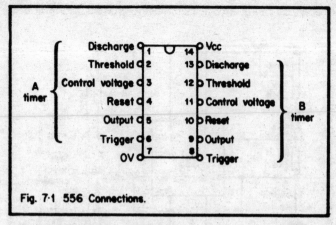

Fig. 7·1 556 Connections.

The device is ideally suited to the many circuits needing two
555s such as the tone burst generator, or the two timer astable.

7:2. 558 and 559 Quad Timers

After the 556 dual timer it will be no surprise to learn that a
16 pin D.I.L. pack is available with four timers. This is done at
the expense of some of the facilities of the 555 and 556.

Firstly the timers do not share the ability of 555 and 556 to
source or sink current. The 558 timers can *sink* 100mA
(0·1A) and the 559 can *source* 100mA (0·1A).

Secondly, no reset line is provided, there not being enough
pins available on the D.I.L. pack.

Thirdly, only one control voltage is available, and this is
common to all the four timers. One simplification is that the
control voltage has been set such that:—

Timed Period T = RC   (rather than 1·1RC)

Finally, each timer can only be used as a monostable, as there

126

is only one comparator in each timer (see internal connections on Fig. 7·2).

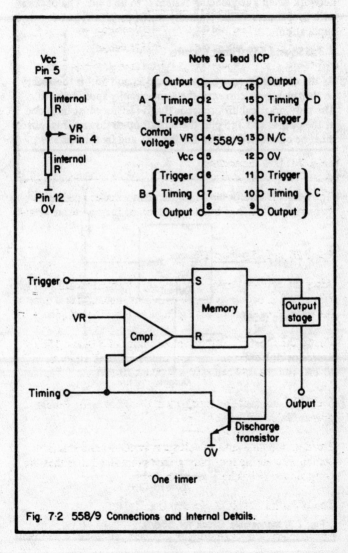

Fig. 7·2 558/9 Connections and Internal Details.

One great advantage has been added, however, the timers can be edge triggered dispensing for the need for a coupling RCD network when a sequencing system is to be built. The package with four timers makes it ideally suited for this type of application.

7:3. 558 and 559 Timer Circuits

As the only difference between the 558 and 559 is the ability to sink or source currents, the same circuits apply to both. The only practical difference is that a resistor must be added at the output (RL) to pull the output up or down. This resistor must be connected to Vcc for the 558 and 0v for the 559.

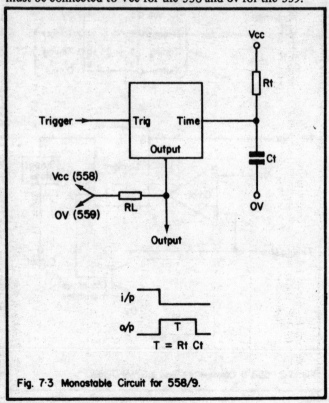

Fig. 7·3 Monostable Circuit for 558/9.

The monostable and astable circuits are shown on Figs. 7·3 and 7·4 respectively. Note that two "quarters" of the device are used to form the astable and that edge triggering allows direct coupling without the need for a coupling resistor capacitor network.

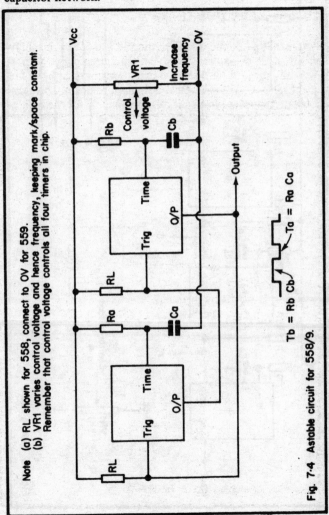

Note (a) RL shown for 558, connect to 0V for 559.
(b) VR1 varies control voltage and hence frequency, keeping mark/space constant.
Remember that control voltage controls all four timers in chip.

$Tb = Rb\ Cb$ $Ta = Ra\ Ca$

Fig. 7·4 Astable circuit for 558/9.

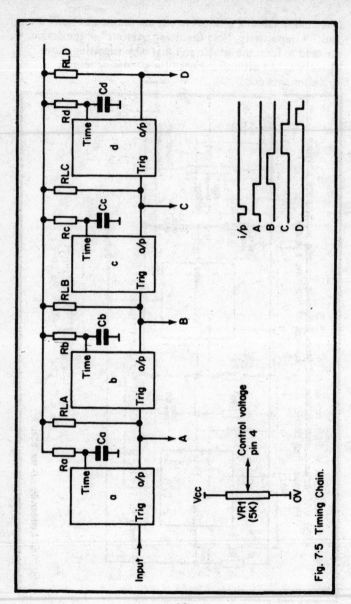

Fig. 7·5 Timing Chain.

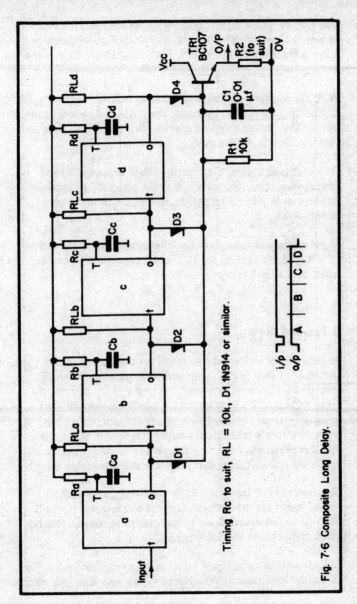

Timing Rc to suit, RL = 10k, D1 1N914 or similar.

Fig. 7·6 Composite Long Delay.

The facility to use direct coupling makes construction of a sequence chain simplicity itself. A typical chain with timing waveforms is shown in Fig. 7.5. Each monostable triggers the next.

With the addition of diodes and emitter followers as in Fig.7.6 composite outputs can be made. The $0.01uF$ capacitor at the base of the emitter follower prevents the brief "glitch" as one monostable triggers the next.

As mentioned before, the control voltage is common to all four monostables. This means that the period of all monostables can be adjusted together, and all the timers remain proportional.

If pin 4 (control voltage) in fig 7.5 is connected to the slider of VR1, adjustment of Ta, Tb, Tc, Td in proportion can be made over a 50:1 range.

8. Practical Notes

In general, the easiest way to construct the circuits in this book is to use 0.1 inch pitch Veroboard. It is quite easy to design layouts for the circuits, a typical example in the metronome from section 6:8, a vero layout for which is shown in Fig.8:1.

Providing care is taken not to bridge tracks with the solder splashes, Veroboard is a very simple way to build circuits. It is readily available from many sources and reasonably priced.

Ready made P.C.B.s are available from firms such as R.S. Components for the basic monostable and astable circuits. With these construction is a simple case of putting in suitable value components and soldering up.

Calculations of component values can present difficulties. The operation requires manipulation of values with K or Meg ohms

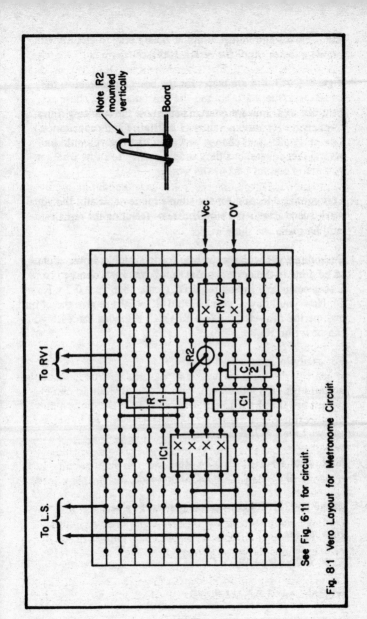

Note R2 mounted vertically

Board

Vcc

0V

RV2

To RV1

R2

R-1-

C1

C2

IC1

To L.S.

See Fig. 6·11 for circuit.

Fig. 8·1 Vero Layout for Metronome Circuit.

and u Farads and milliseconds. It is very easy to end up with results a factor of 10 (or even a 1000) out.

Figs. 8:2, 8:3, 8:4 are tables for the basic monostables, the equal mark/space astable and the general astable. These are read like a car mileage chart. These show the resulting times/frequencies for various values of resistance and capacitance. Obviously all values cannot be shown, the charts would be several feet square, but they should suffice to show whether you are in correct part of the world.

Component selection for the simple monostable and the equal mark/space astable is a simple case of selecting the right values and heigh-ho the thing works.

Designing a particular waveform for the astable, however, has to be done in the correct order as follows. First choose R and C to give the low period required, (remember Tb = 0.7 x Rb x C). Now you know C you can find (Ra + Rb) to give the output high period (Ta = 0.7 x (Ra + Rb) x C) Knowing Ra + Rb you can now find Ra.

For example, suppose we want;

output high (Ta) = 0.9 sec.
output low (Tb) = 0.1 sec.

and we have a $4.7u$F capacitor.

First select Tb = 0.7 x Rb x C
         0.1 = 0.7 x Rb x $4.7u$F

gives Rb = 33K for the nearest preferred value.

Now we use;

Ta = 0.7 x (Ra + Rb) x C

$0.9 = 0.7$ x (Ra + Rb) x $4.7u$F

Timing Capacitor

Timing Resistor	0·001uF	0·01uF	0·1uF	1uF	10uF	100uF	1000uF
1K0	—	11uS	110uS	1·1mS	11mS	110mS	1·1S
10K	11uS	110uS	1·1mS	11mS	110mS	1·1S	11S
100K	110uS	1·1mS	11mS	110mS	1·1S	11S	110S
IMO	1·1mS	11mS	110mS	1·1S	11S	110S	1100S

Fig. 8:2. Table for 555 Monostable (T = 1·1 RC Secs)

Timing Capacitor

Timing Resistor	0·001µF	0·01µF	0·1µF	1µF	10µF	100µF	1000µF
1K0	—	70KHz	7KHz	700Hz	70Hz	7Hz	1·4S
10K	70KHz	7KHz	700Hz	70Hz	7Hz	1·4S	14S
100K	7KHz	700Hz	70Hz	7Hz	1·4S	14S	140S
1M0	700Hz	70Hz	7Hz	1·4S	14S	140S	1400S

Freq.   Period

Fig. 8:3 Table for Equal Mark/Space Astable (Period = 1·4RC, Freq. = $\frac{1}{1·4RC}$)

Timing Capacitor

Timing Resistor	0·001uF	0·01uF	0·1uF	1uF	10uF	100uF	1000uF
1KO	—	7uS	70uS	700uS	7mS	70mS	700mS
10K	7uS	70uS	700uS	7mS	70mS	700mS	7S
100K	70uS	700uS	7mS	70mS	700mS	7S	70S
1MO	700uS	7mS	70mS	700mS	7S	70S	700S

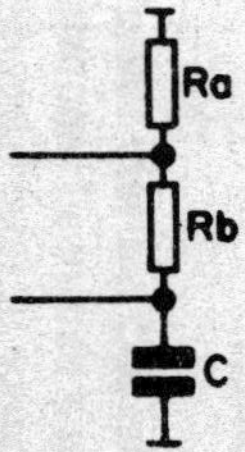

Output Low $= 0.7\, RbC$

Output High $= 0.7\,(Ra + Rb)\,C$

Total Period $= 0.7\,(Ra + 2Rb)\,C$

Fig. 8:4. 555 Astable Table of 0.7 RC

gives $(Ra + Rb) = 270K$

since $Rb = 33K$

$Ra = 220K$ to the nearest preferred value.

The values are then;

Ra   =   220K
Rb   =   33K
C   =   4·7uF

Usually component selection is based on what is available in the "come in handy box". There are a few points worth noting however. For long periods it is better to use a large value capacitor and small value resistor than the other way round, as this gives a lower impedance at the RC junctions and hence better noise immunity. Unfortunately, large capacitors are expensive.

Electrolytics are troublesome things. They have a very large tolerance and a large leakage current. In addition they do not become capacitors proper until about 10% of their rated voltage Periods calculated with electrolytic capacitors can be greatly different from what you actually get.

If it is desired to trim the period with an external variable resistor it is preferable to do this on pin 5 (see section 2.9). If this cannot be done, mount an end stop resistor local to the chip at the capacitor end (see Fig. 8:5). This resistor acts as a filter to reduce any noise induced in the lead from the variable resistor.

The 555 is a very tolerant device to electrical noise, but by their nature all monostables can be noise prone. To reduce the chance of false triggering or unpredictable periods, the following precautions should be taken:—

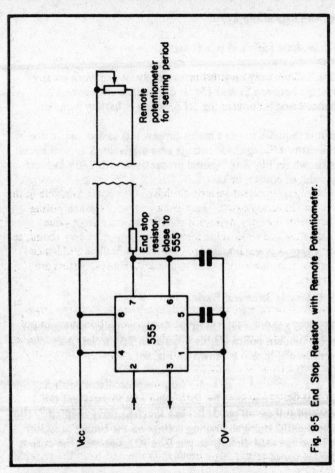

Fig. 8-5 End Stop Resistor with Remote Potentiometer.

a). Keep timing components and trigger leads short.

b). De-couple pin 5.

c). Use large C and small R.

d). Never run the connection to pin 2 through a switch or relay contact.

e). Use a good supply.

f). Separate logic and power earths.

The 555 is a very easy device to supply. It can work on any voltage between 5v and 15v. It draws little current and is hence suitable for running off 6, 9 or 12v battery supplies.

If it is required to use a mains supply, it is easiest to use one of the many I.C. regulator circuits now available. A typical circuit is shown on Fig. 8:6. Normal precautions with 240V circuits should, of course, be taken.

9. Additional Noise Makers

9:1. Simple Siren and Warble

Fig. 6:6 gives one version of an American police siren and the more modern police warble popular in T.V. series. A simpler version of Fig. 6:6 is shown on Fig. 9:1.

IC2 is connected as an audio frequency oscillator, with its frequency determined by the timing components and the control voltage on pin 5. IC1 is a low frequency oscillator. The exponential rising and falling voltage on the timing capacitor C1 is connected directly to pin 5 on IC2, sweeping the output frequency of IC2.

Resistor R2 determines the period of the ascending frequency, R1 the period of the descending frequency. It is quite a flexible circuit as these can be adjusted independently to produce some very interesting sounds. By lowering the value of C1 as shown on Fig. 9:1 the siren is speeded up to a warble.

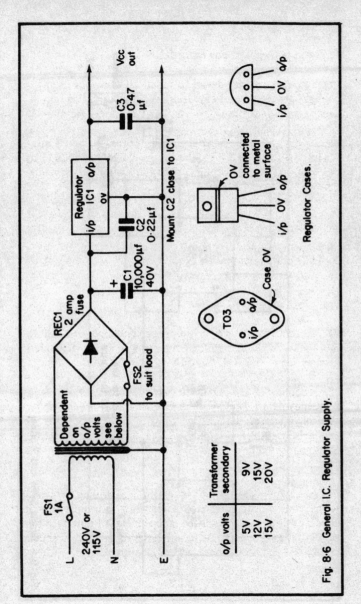

Fig. 8·6 General I.C. Regulator Supply.

141

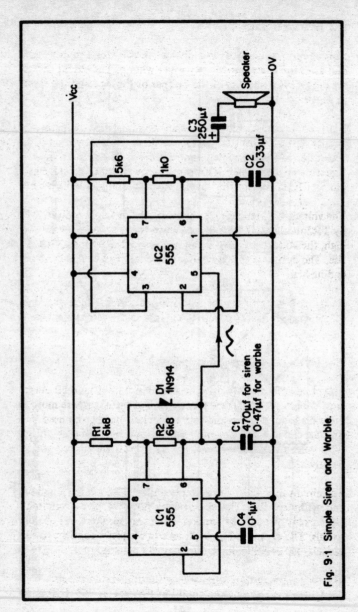

Fig. 9·1 Simple Siren and Warble.

142

9:2. Destroyer Siren

This circuit was designed as a distinctive siren for a motor boat, and gives the characteristic destroyer whoop. The circuit shown on Fig. 9:2 is a mixture of the bleeper of Fig. 6:4 and the siren in Fig. 9:1.

IC1 is connected as a low frequency asymmetrical oscillator, and its output is inverted by TR1 to drive the reset pin on IC2. The latter 555 is connected as an audio oscillator, and it is therefore enabled when IC1 output is low. With the voltage at pin 5 of IC2 constant the output of the circuit would be bleeps.

The voltage on capacitor C1, however, is fed to pin 5 on IC2 by TR2 modulating the output. When the reset on pin 4 is high, the voltage on pin 5 is falling causing the frequency to rise. The output is thus a whoop sound, starting low and ending high.

Resistor R1 determines the repetition rate and R2 the duration of the whoop. Resistors R3/R4 determine the centre frequency.

9:3. Driving Low Impedance Speakers

The basic 555 can only drive an impedance of around 50 ohms, and 75 ohm speakers were recommended earlier. Where more noise is required lower impedance speakers need to be used. The circuits in Fig. 9:3a and b show simple ways of doing this. With these circuits speakers as low as 4 ohms impedance can be driven.

The circuit in Fig. 9:3a is a simple buffer. TR1 being turned on and off by the 555. This circuit does, however, give a standing DC current through the speaker, and some speakers will object to this. The circuit in Fig. 9:3b uses two emitter followers and capacitance coupling to remove any D.C. component.

The maximum output attainable without using an output transformer is given by the circuit in Fig. 9:4. A 555, IC1, is

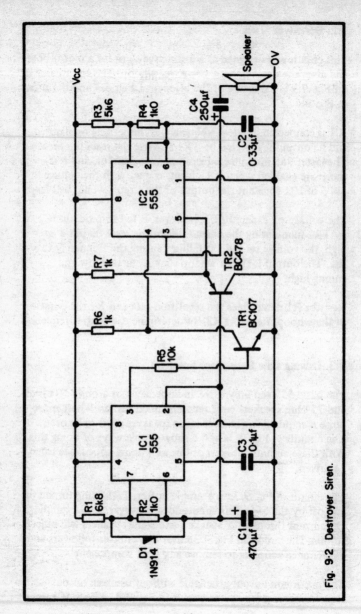

Fig. 9-2 Destroyer Siren.

144

used purely as an inverting buffer and four emitter followers are used. The peak to peak voltage across the speaker is now double the supply voltage. This bridge circuit needs no output coupling capacitor as there is no standing D.C. component through the speaker.

Where very loud speakers are required, external audio amplifiers or P.A. amplifiers can be used to give any required wattage. In all cases it should be remembered that the speaker currents will be very high, and the precautions outlined in Section 2:5 should be followed.

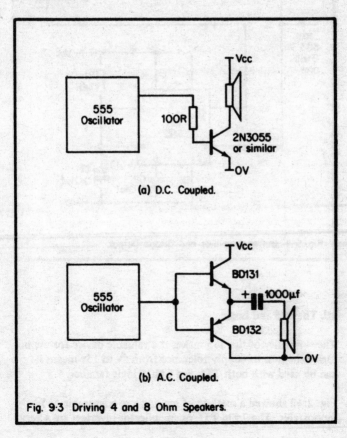

Fig. 9·3 Driving 4 and 8 Ohm Speakers.

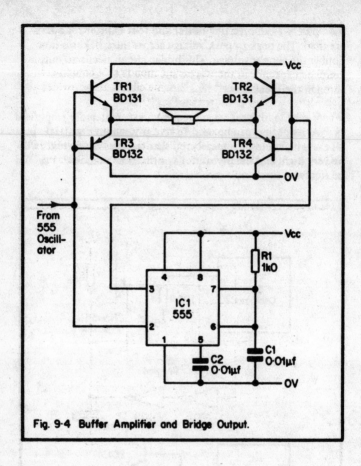

Fig. 9·4 Buffer Amplifier and Bridge Output.

10. The 555 and Logic

The versatility of the 555 makes it a valuable device for use in logic schemes. Its supply tolerance from 5v to 15v means it can be used with both TTL and CMOS logic families.

Fig. 2:12 showed a method of making a re-triggerable 555 monostable. The 7406 TTL open collector inverters are a very useful device for implementing this circuit as part of a logic scheme.

Fig. 10:1 shows a 555 re-triggerable monostable constructed using the 7406 hex inverter. The same circuit also provides a delay 0 − 1 circuit as shown on Fig. 10:2.

If the inverter is put onto the input as shown on Fig. 10:3 we can get a delay 1 − 0 circuit. The delay circuits in Fig. 10:2 and 10:3 are particularly useful as switch bounce filters on inputs from single pole switches.

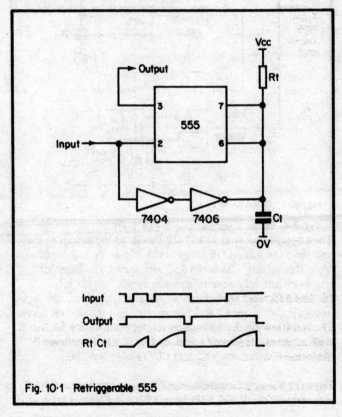

Fig. 10·1 Retriggerable 555

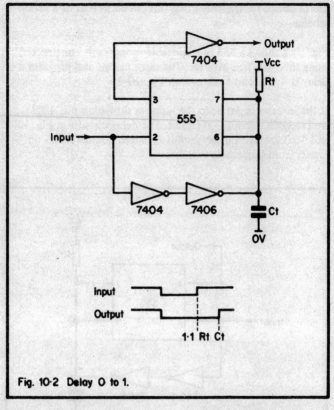

Fig. 10·2 Delay 0 to 1.

The trigger input (pin 2) and the threshold input (pin 6) set and reset the internal flip flop in the 555 at ⅓ and ⅔rds of Vcc. This facility makes the 555 very useful as a Schmidt trigger circuit for converting slowly varying inputs into digital levels. Fig. 10:4 shows a typical arrangement, the input being applied to pins 2 and 6 in parallel and the output taken as usual from pin 3. The high input impedance on pins·2 and 6 allow this arrangement to be used with high impedance devices.

Edge triggering is frequently used in logic, and the circuit of Fig. 2:5 will work well with both TTL and CMOS.

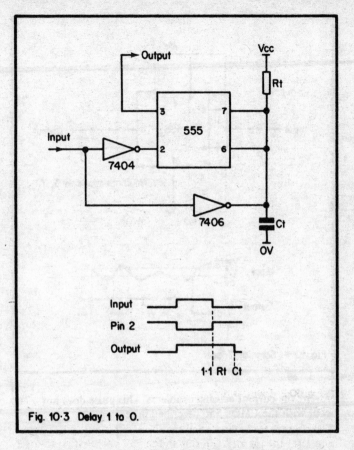

Fig. 10·3 Delay 1 to 0.

The output drive capability of the 555 gives excellent fan out, and it can drive far more gates in both CMOS and TTL than will be required in any system. The actual fan out to TTL is well in excess of 50 and is almost infinite for CMOS.

The 555 and 556 have a slightly antisocial habit of drawing a large current pulse from the supply as they change state. This pulse causes voltage spikes which can cause problems elsewhere in the logic. A 0.01µF capacitor should therefore be placed between pins 8 and 1 on a 555 (pins 14 and 7 on a 556) to

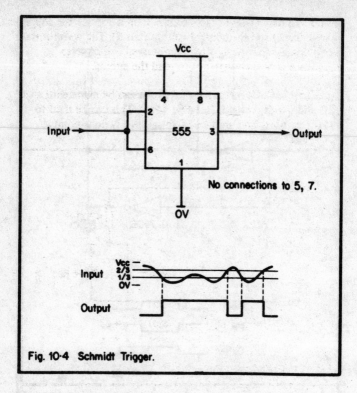

Fig. 10·4 Schmidt Trigger.

prevent the current causing problems. This pulse does not
occur with the CMOS 555 described in section 12.

11. Fault Finding

You have built your 555 circuit, and it does not work, what
do you do next? Without a doubt, the commonest faults are
constructional faults and a visual inspection will often pick up
solder splashes, uncut Vero tracks and shorting leads.

To proceed further you really need a multimeter, and after a
soldering iron and snips the meter must rank top of the
amateurs needs. Voltage checks around a circuit will often reveal

obscure faults. Obvious places to check on a 555 are the supply-pins (8 and 1) and the control voltage (pin 5). The waveform on the timing capacitor can often be observed if the meter impedance is high enough not to load the circuit.

In the absence of a meter a useful probe can be made with an LED and a series resistor (see Fig. 11:1). This can be used to test for presence and absence of voltages at various points.

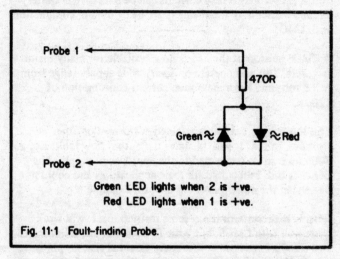

Green LED lights when 2 is +ve.
Red LED lights when 1 is +ve.

Fig. 11·1 Fault–finding Probe.

High speed circuits (such as the dice) cannot be tested with a meter. A useful trick here is to slow the circuit down by changing the timing components. If C2 in Fig. 5:9 is increased to, say, 1000 μF, the circuit will operate slowly allowing its operation to be monitored. The sound making circuits can also be de-bugged in this manner.

Circuits using electrolytic capacitors can often operate in a peculiar manner. The wide tolerances can cause circuits to operate faster or slower than expected. Normally this causes no problem, but some circuits such as the car emergency flashers (section 3:4) or the siren circuits may need trimming to give the right rate. Usually altering the value of one of the timing resistors slightly is all that is required.

The 555 itself is a very robust device, and will usually survive wiring errors. As a general rule it only fails if the output is shorted to a supply rail for an extended period.

12. Further 555 Variations

The 555 draws about 7mA off the supply, and the 556 about 10mA (at 9 volts). These currents can be excessive in battery circuits particularly if the timers are being used in conjunction with CMOS.

A CMOS version of the 555 is now available for many manufacturers. This will operate on a very wide supply range from 2 to 18 volts and has a maximum current consumption of 120µA.

This low supply current is attained at the cost of other facilities. In the '1' and '0' states the output now looks like a 400ohm resistor connected to the appropriate supply. The device is thus best suited for logic applications and could not be used in the noise maker circuits, for example.

Even in logic systems it has some restrictions. It will work quite well with CMOS, but with TTL the fan out is limited to two standard loads. It is absolutely essential that no pin is taken outside the supply rails, or the device will probably fail.

The output stage of the CMOS 555 does not produce current pulses on the supply (see section 10) and decoupling is not necessary.

The CMOS 555 is pin compatible with the standard 555. A CMOS pin compatible 556 is also available with similar characteristics.

An interesting 555 called the 355 is available from Teledyne. It is in all respects equivalent to a normal 555 (including supply current and drive capability). It is, however, guaranteed free of any current pulse when changing state. It is a somewhat expensive device, but is very useful in logic schemes where the full 555 drive specification is needed.

13. Further General Circuits

13.1 Microprocessor Related Circuits

The 555 family is to be found in many microcomputers; common applications being clock generators for serial UARTs as Fig. 13.1 and the watchdog timer describer earlier in section 2.6.

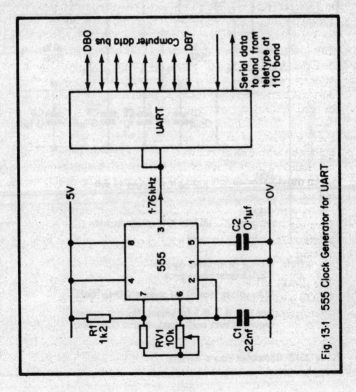

Fig. 13·1 555 Clock Generator for UART

The 555 can also be connected to a computer parallel port to provide interesting effects. Most computer games are improved by the addition of sound effects, and Fig. 13.2 shows three 556 used to provide the blips, pings and buzzes for a typical game. The 'a' portions are used as a monostable to gate the oscillator 'b' portions. The three outputs are summed by IC4

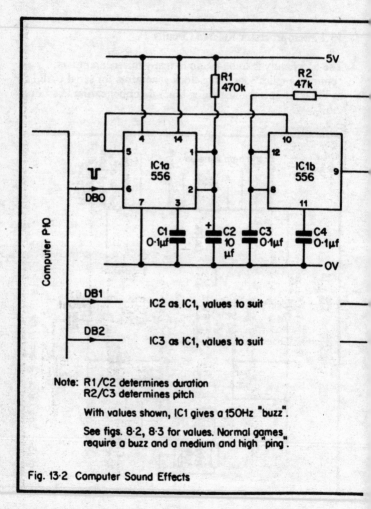

Fig. 13·2 Computer Sound Effects

Note: R1/C2 determines duration
R2/C3 determines pitch

With values shown, IC1 gives a 150Hz "buzz".

See figs. 8·2, 8·3 for values. Normal games require a buzz and a medium and high "ping".

to drive a small speaker. (On most small computers the ±12 volt supply for the RAM is not adequate to drive a speaker, and an additional ±12 volt supply needs to be added for IC4). The software needed to drive Fig. 13.2 is minimal, 1−0−1 is simply written to the corresponding bit when a tone burst is required.

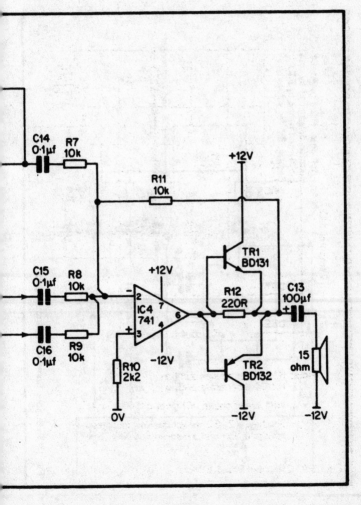

A simpler circuit, but one requiring more programming effort, is shown on Fig. 13.3. The port outputs are connected to the reset pins on the oscillators IC1-IC3. These will oscillate (and produce a sound) as long as the reset is held at a '1'. The tone duration thus has to be determined by the program, whereas in Fig. 13.2 it was done by the 'a' portion of the 556.

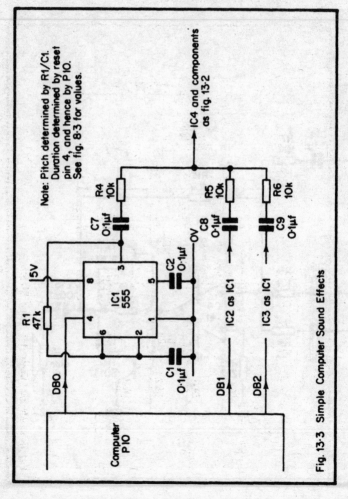

Fig. 13-3 Simple Computer Sound Effects

156

In Fig. 13.4 the port outputs are connected by resistors to the control pin 5 of a single 555 oscillator, with bit 0 being connected to the reset. The bit pattern written to bits 1 to 4 will alter the period as explained in section 2.9. The reset from bit 0 is used to determine the duration. This circuit again requires a fair amount of programming effort, and cannot produce two simultaneous sounds.

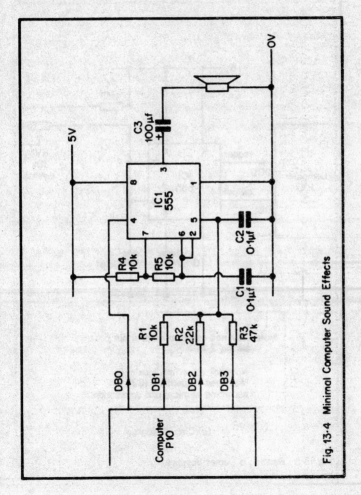

Fig. 13-4 Minimal Computer Sound Effects

157

Joysticks are a popular addition for many games, and Fig. 13.5 shows how a 555 can be used to read a joystick position. The 555 is connected as a monostable, with period determined by the joystick position. When the joystick position is needed by the computer, bit '0' is used to trigger the monostable, and

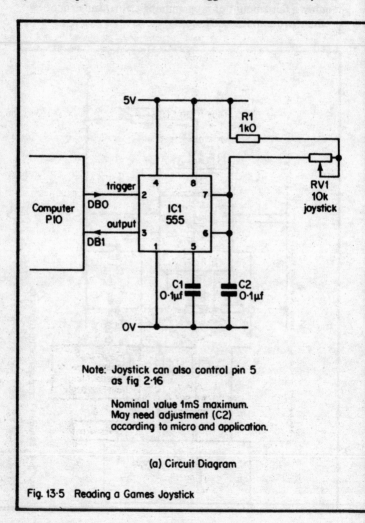

Note: Joystick can also control pin 5 as fig 2·16

Nominal value 1mS maximum. May need adjustment (C2) according to micro and application.

(a) Circuit Diagram

Fig. 13·5   Reading a Games Joystick

the output duration (and hence the joystick position) determined by the flow chart as Fig. 13.5b. The flow chart should be written as a subroutine which is called whenever a joystick position is needed.

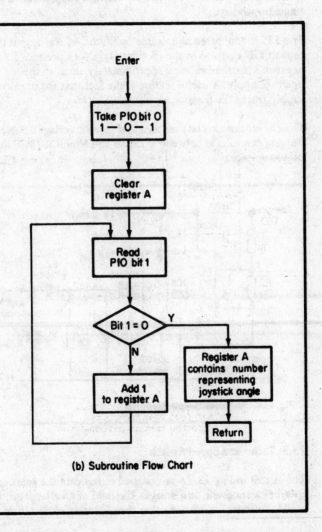

(b) Subroutine Flow Chart

13.2. Single to Dual Rail Supply

Operational amplifiers, such as the 741, need both a positive
and a negative supply. Fig. 13.6 shows a simple circuit to
provide a low current ($\sim$ 50mA) negative supply from a
single positive rail.

The 555 is connected as a simple oscillator, whose output is
capacitively coupled to the rectifier D1, D2 to produce a
negative smoothed voltage, approximately equal to the
positive supply. A useful feature is the fact that the negative
supply tracks the positive supply.

The circuit can also be used as a low current voltage doubler
because the voltage between V minus and V plus is twice the
supply voltage.

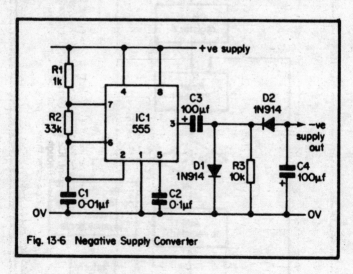

Fig. 13-6 Negative Supply Converter

13.3. Tacho and Speed Switch

The circuit in Fig. 13.7 was designed to indicate the rotational
speed of a windmill, and bring in the field of an alternator. It
could, however, be used in other speed sensing applications.

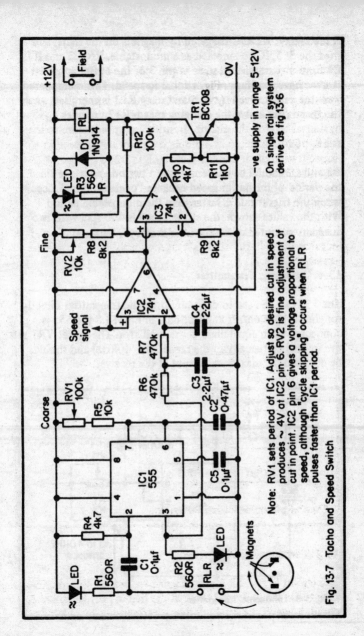

Note: RV1 sets period of IC1. Adjust so desired cut in speed produces ~ 4V at IC2 pin 6. RV2 is fine adjustment of cut in point. IC2 pin 6 gives a voltage proportional to speed, although "cycle skipping" occurs when RLR pulses faster than IC1 period.

Fig. 13·7 Tacho and Speed Switch

161

A reed relay, RLR, is triggered by magnets on the shaft, and fires the 555, IC1 connected as a monostable. R6, C3 and R7, C4 form a two stage filter, so at pin 3 of the buffer amplifier IC2 we have a voltage proportional to speed. This is compared with the preset speed by IC3, and relay RL1 is energised when the speed rises above the charging speed. R12 provides hysterises to stop the circuit "hunting" at speeds close to the cut in point.

R5 and C2 should be chosen to give a period about one third the period of the cut in speed because "cycle skipping" occurs when the trigger rate is faster than the monostable period. With the values shown, the cut in speed is 400 rpm with two magnets mounted at 180° on the shaft (for balance).

13.4. Ultrasonic Transmitter

The 555 can be used to drive an ultrasonic transmitter directly for simple tone on/off applications. In Fig. 13.8 the 555 is connected as an equal mark space oscillator. The preset VR1 sets the frequency (usually in the range 30 – 40kHz) and should be adjusted for maximum output at the receiver.

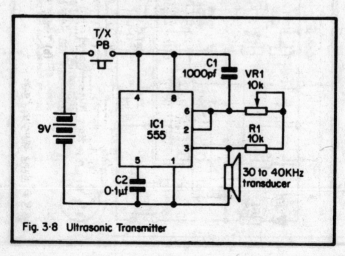

Fig. 3·8 Ultrasonic Transmitter

13.5. Pulse Stretcher Probe

Fig. 11.1 showed a simple logic probe, albeit with many short-comings. Fig. 13.9 shows a high input impedance probe that not only indicates '1's and '0's, but can also show the prescence of pulse trains and indicate the presence of short duration pulses.

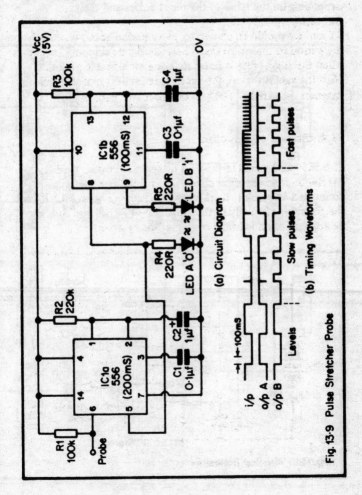

Fig. 13-9 Pulse Stretcher Probe

IC1 is a 556 connected as two monostables of period 0.2 and 0.1 secs. IC1a drives the '0' LED A, and IC1b the '1' LED B. The circuit operates as shown on Fig. 13.9b, and will indicate pulse trains by the flashing of LEDs A and B due to the pulse stretching. For frequencies below about 5Hz, the pulses can be observed directly. For frequencies above 5Hz, both LEDs will appear on, or A will be on and B blinking at 5Hz depending on the sense of the input pulses and their frequency.

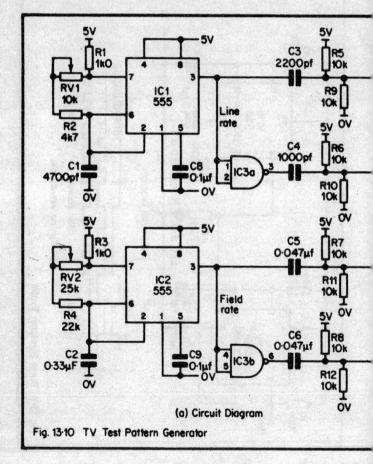

(a) Circuit Diagram

Fig. 13·10 TV Test Pattern Generator

13.6. TV Test Pattern Generator

This circuit generates a simple cross pattern of Fig. 13.10b for testing CCTV monitors and computer VDUs. The circuit, shown on Fig. 13.10a uses two 555 oscillators running at field rate (50Hz) and line rate (15kHz) to produce line and field sync and bright up pulses for RC differentiators. These are summed by IC4a and b to give a composite 1v p to p video test circuit.

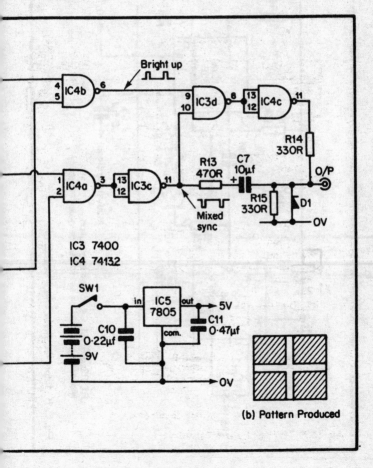

IC3 7400
IC4 74132

(b) Pattern Produced

The circuit was designed to run on a 9 volt battery, so a 78L05 regulator is used to provide the 5 volt supply used by the circuit. If required, an ASTEC UHF modulator could be added to test domestic televisions.

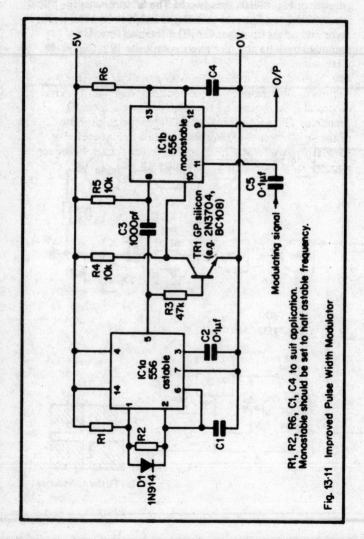

R1, R2, R6, C1, C4 to suit application.
Monostable should be set to half astable frequency.

Fig. 13·11 Improved Pulse Width Modulator

13.7. Improved Pulse Width Modulator

A simple pulse width modulator was described in section 2.10.
The circuit on Fig. 13.11 gives improved performance at the
expense of additional components. The 'a' section of the 556
is connected as an astable with D1 included so that the mark/
space ratio is greater than 50%. The charging time is
determined by R1, C1, the discharge time by R2, C1.

The 'b' portion is a monostable with period set by R6, C4
and the modulated signal to the control voltage pin 11. The
monostable is triggered by the astable 'a' portion via C2.
Transistor TR1 resets the monostable, so the pulse width
of the monostable cannot exceed the low output period of
the astable waveform. The circuit therefore does not produce
unpredictable outputs if overmodulated.

Notes

Notes

Notes

Notes

Notes